WONDERGENES

MAXWELL J. MEHLMAN

WONDERGENES

GENETIC ENHANCEMENT
AND THE FUTURE OF SOCIETY

INDIANA
University Press
Bloomington & Indianapolis

This book is a publication of

Indiana University Press
601 North Morton Street
Bloomington, IN 47404-3797 USA

http://iupress.indiana.edu

Telephone orders 800-842-6796
Fax orders 812-855-7931
Orders by e-mail iuporder@indiana.edu

The paper used in this publication meets the
minimum requirements of American National
Standard for Information Sciences—Permanence of
Paper for Printed Library Materials, ANSI Z39.48-
1984.

Manufactured in the United States of America

Library of Congress Cataloging-in-Publication Data

Mehlman, Maxwell J.
  Wondergenes : genetic enhancement and the future
of society / Maxwell J. Mehlman.
    p. cm. — (Medical ethics series)
Includes bibliographical references and index.
  ISBN 0-253-34274-0 (cloth : alk. paper)
  1. Genetic engineering—Social aspects. 2. Medical
genetics—Social aspects. I. Title. II. Series.
  QH438.7.M44 2003
  303.48'3—dc21
                                        2002156519

1 2 3 4 5 08 07 06 05 04 03

To Cher

Contents

Acknowledgments

The research for this book was supported by a grant from the Ethical, Legal, and Social Implications Program of the National Human Genome Research Institute (R1HG01446B2). The author wishes to thank Emily E. Raemore and Daniel Fishbein for their excellent research help, his co-investigators Eric Juengst and Thomas Murray for their input on earlier stages of the project, Nancy Pratt Kantor, Drew Bryan, Jane Lyle, and Marilyn Grobschmidt at the Indiana University Press, and his family.

WONDERGENES

The call comes into the Ski Patrol at dusk. "Family reports their eleven-year-old son didn't show up at the end of the day. His friends say he went out of bounds on his snowboard, heading over toward Crystal Cliffs." The rescue hut bursts into action. The two patrollers assemble their equipment, don heavy clothing, insert communications buttons in their ears, and radio Mountain Rescue for special personnel and a chopper. The last thing they do before grabbing their skis and heading out into the swirling snowstorm is to swallow "boosters"—capsules containing high doses of short-acting, genetically engineered drugs. These drugs are available only to law enforcement, emergency personnel, and others with a special license from the government. Within minutes, the patrollers' cognitive functioning, physical dexterity, hearing and visual acuity, and physical strength and stamina are increased an average of forty percent.

Minutes later they stand atop the cliff face. Staring down, they spot the boy lying on a narrow rock ledge far below them. They radio the helicopter, which homes in on their position and lowers one of its occupants onto the ledge. This individual, a professional mountain rescuer, has undergone genetic engineering to install genes that give her special abilities. Clinging with her specially developed fingers to the icy wall, she quickly diagnoses the child as having a number of life-threatening injuries. The wind is howling, and it would be too dangerous to try to winch the boy back up into the hovering aircraft. So the rescuer bundles him into a collapsible litter and hauls it single-handedly up the sheer face of the cliff. The entire operation takes less than fifteen minutes and saves the child's life.

How realistic is this scenario? Right now it is science fiction. But the chances of it occurring may be greater than you think. The Human Genome Project—the government program to map and sequence human DNA—is almost complete, and work is underway to identify the genes that code for the myriad of inherited human characteristics. At the same time, we are learning how to manipulate these genes—multiplying or canceling their effects or installing entirely new capabilities in an individual's DNA. Genetically engineered drugs already are widely available, and new ones being tested include drugs to improve mental functioning. Aimed at treating Alzheimer's disease, these drugs may prove able to enhance cognition in "normal" individuals as well. Genetic tests soon will be developed that can identify embryos or fetuses not just with genetic abnormalities, but with desirable physical or mental traits, enabling parents to decide which ones to implant in the womb or bring to term. Researchers are beginning to experiment with altering human DNA. While the initial efforts are mainly aimed at treating or preventing disease, the same approaches could be used to improve non-disease traits.

To be sure, there are many technical hurdles to overcome before we can create genetically enhanced individuals like the ones in the rescue scenario. For example, the Human Genome Project recently has revealed that there are far fewer human genes than expected, which will make it much more difficult to determine how they interact with each other and with the environment to produce human traits. But while this may slow the development of methods for altering human characteristics, it is unlikely to derail them. As one leading geneticist declares: "There are no insurmountable scientific barriers to genetic enhancement."[1]

The reason is simple but profound. Every day we are witnessing revolutionary breakthroughs that are enabling us to sequence the building blocks of life and begin to understand the mechanisms that regulate them. But the explosion in the science of human genetics is the result not of just one revolution but of two. Coinciding with the revolution in biology is the revolution in computers. Recently, a private company, Celera, accomplished in less than a year what it took a government research consortium ten years to achieve: It by and large sequenced the human genetic code. Celera was able to do this by using not only state-of-the-art molecular science, but the power of supercomputers. In fact, at the time, the Celera project represented the largest private use of supercomputing capacity in the world.

A confluence of transformational forces of this magnitude is unprecedented in human history. Think back to little more than twenty years ago. You were lucky if your personal computer possessed sixty-four *kilo*bytes of memory, and even at that, the machines were few and far between. In the space of two decades, the power of the computer has increased exponentially, and we now employ them in virtually all aspects of our lives. Now go back only fifty years, to when the physical structure of DNA was first discovered, and neither modern genetics nor modern computers were more than a gleam in the visionary's eye. How far we have come.

Together, the twin revolutions in biology and computers already are producing enormous social good. Drugs made with recombinant DNA have relieved shortages and made life-saving therapies widely available. Genetic testing can help diagnose disease and forewarn individuals so they can take preventive measures. Scientists are on the brink of successfully modifying genes to cure inherited disorders.

But another prospect looms: the ability to genetically enhance human beings, to give them capabilities hitherto rare or even impossible. Once again, the benefits could be enormous. Enhanced medical researchers could more rapidly find cures for disease. Pilots and soldiers could gain better vision and greater physical and mental acuity. Explorers could survive in hostile environments like under the sea or in outer space. Rescuers could save lives.

But there is a darker side to genetic enhancement. Consider the

Harvard Business School Class of 2042. Applicants are selected on the basis of their genetic profiles. Only the "right" ones are accepted, based on their enhanced ability to master the curriculum, become captains of industry, and enrich the school with their donations. "Natural" kids don't stand a chance; the class is filled with the offspring of wealthy parents who were able to afford the latest enhancement technologies provided as part of the in vitro fertilization (IVF) process. Having received their genetic advantages at an early stage of life, these children will pass them on to their children and their children's children, assuming they marry well and make the correct genetic decisions. The new genobility quickly secures its grip on the major financial, social, and political institutions of the country. Although all adults still get a vote at election time, only enhanced candidates win.

The point is that the same technological breakthroughs that make dramatic social benefits possible also threaten to undermine the foundations of our society. Ultimately, the social forces unleashed by these wondergenes could unravel the fabric of society itself, plunging the world into a new Dark Age of feudal tyranny and civil unrest.

Can we avoid this nightmare future? To do so, will we have to forgo the benefits of genetic enhancement? Or is there a way to capture the benefits without unleashing such destructive forces? And if there is, will we act in time?

1

An Announcement at the White House

"Nearly two centuries ago, in this room, on this floor, Thomas Jefferson and a trusted aide spread out a magnificent map—a map Jefferson had long prayed he would get to see in his lifetime. The aide was Meriwether Lewis and the map was the product of his courageous expedition across the American frontier, all the way to the Pacific. It was a map that defined the contours and forever expanded the frontiers of our continent and imagination.

"Today, the world is joining us here in the East Room to behold a map of even greater significance. We are here to celebrate the completion of the first survey of the entire human genome. Without a doubt, this is the most important, most wondrous map ever produced by humankind."

With these remarks, made before a White House audience on June 26, 2000, President Clinton announced the successful sequencing of the human genome.

As he spoke, the president was flanked by two men. One of them, Francis Collins, had inherited in 1984 the mammoth government decoding program, known as the Human Genome Project, from James Watson, co-discoverer in 1953 of the double-helix shape of the DNA molecule. Since then, Collins had overseen one of the largest non-defense research programs in history. This moment was the crowning achievement of his career. He had already won fame for discovering the gene for cystic fibrosis in 1989 and for collaborating in the discovery of the Huntington disease gene. Now the president of the United States was giving him credit for successfully completing the Human Genome Project, a science project that, more than any other, could revolutionize life as we know it.

But Collins had reason to be disgruntled. On the other side of the president stood J. Craig Venter, president and chief scientific officer of Celera Genomics. His company also had sequenced the human genome. But unlike the federal project, which had taken nine years and cost $3 billion, Venter's company had achieved roughly the same result in less than nine months, at a cost of only $200 million. It was an astonishing achievement by any account, and at the White House, it robbed Collins of center stage.

To be sure, Celera had not started its sequencing program from scratch. Other scientists had perfected some of the techniques that Celera had employed, like recombining DNA, and Celera had access to the results of the government project, which publicly posted its newly discovered sequences on the Internet. But Celera had been founded by visionaries, and they employed many clever expedients. Instead of using semi-automated sequencing machines, which required hand loading and unloading, Venter employed fully automated machines developed by a company called PE, which had changed its name from Perkin Elmer after defects in its mirror had crippled the Hubble space telescope. Celera had used a lot of these automated machines, three hundred of them in fact. And they had run them around the clock.

The cleverest stratagem of all was the way Celera used computers to analyze the data that the sequencing machines produced. At the time, as noted in the Introduction, Celera was employing the largest non-governmental supercomputing capacity in the world.

When Celera first started up, some of the scientists working on the government's Human Genome Project had been skeptical and disparaging.[1] They said that Celera's "shotgun approach"—another of its innovations, which enabled the order of small fragments of

DNA to be determined rapidly—would not work. Celera might be able to decode the DNA sequences, but it would not be able to patch all the fragments back together in the proper order. Celera countered that it would not only reassemble the fragments, but would produce a complete sequence sooner than the government.

Francis Collins, as head of the government project, took up the gauntlet. He vowed to complete a "rough draft" of the human genome by 2000 and a final version by 2003, rather than by the original target date of 2005. With this, what had been an intense scientific rivalry became a flat-out horse race. When Venter scoffed that the Human Genome Project would fail to meet this timetable, scientists working on Collins's project reiterated that Celera would not be able to reassemble its fragments into an accurate map. The race heated up even more when the Human Genome Project began cranking out large sequences in keeping with its new schedule. Celera in turn defied its critics by publishing an accurate sequence of the fruit fly genome.

Ari Patrinos decided to intervene. Patrinos headed up a portion of the Human Genome Project run by the Department of Energy. This was a smaller program than Collins's at the NIH. Congress had created it in recognition of the DOE's long-standing interest in genetics, dating back to the days when, as the Atomic Energy Commission, it had begun researching the genetic effects of atomic fallout. Determined to achieve a reconciliation between Collins and Venter, Patrinos invited them to his home for pizza. Several more meetings followed, and the two scientists finally agreed to bury the hatchet and make a joint announcement of success. The ceremony took place in June at the White House. The Human Genome Project later published its sequence map in the journal *Nature,* while Celera published its map in *Science.*

Only once before had President Clinton stood in front of the White House microphones to speak to the nation about a scientific discovery. This had occurred when scientists declared that they had discovered extraterrestrial life in a fragment of a Mars asteroid. While this discovery, if true, would be astounding, the June 2000 announcement that the human genome had been sequenced marked a "sea change" from the standpoint of human biology, one of the few times that the overused label was appropriately applied.

Moreover, the White House genome announcement warranted this characterization even though it was untrue. Neither the Human Genome Project nor Celera actually had completed the sequencing of human DNA. The Human Genome Project had only sequenced 97 percent of the genome and only 85 percent of the sequences had been placed in the proper order. Although Celera claimed it had sequenced 99 percent of the genome, it admitted that it was still piecing the fragments together.

But these were mere details. The important thing was that the sequencing would be completed soon.

What was so important about sequencing the human genome? What does it enable us to do? The answer is not much by itself. It is but one step in a journey, and a relatively initial step at that. But it is a monumental accomplishment nonetheless. The sequence is three billion units long. This is such a large number that science commentators like to use equivalents to describe it, such as a fifteen-foot-high stack of computer printouts, or three full sets of the Encyclopedia Britannica, with each letter standing for a genetic unit. The location and order of this sequence was not completely known by June 2000, but most of it was.

Aside from the sheer size of the achievement, sequencing the human genome is important because the information it reveals is so fundamental. Consider its closest analogy—the periodic table of elements that hung on the wall of your high school chemistry lab. By itself, the table of elements does not do much. But imagine modern science without the knowledge it contains about the basic building blocks of chemistry and physics. Sequencing the human genome serves the same function for human biology. It is a catalogue of the building blocks of life.

But the significance of sequencing the genome goes far beyond serving as a source of basic scientific knowledge. Its significance lies in its trajectory. It is a step, a crucial step at that, on the path to an unprecedented evolutionary destination. At the end lies nothing less than the ability to genetically alter the human species.

The skeptic leaps up. "This is completely ridiculous!" he cries. "Genetic science is far more complicated than that. The complex traits that make up what it means to be human are the product of multiple genes, and of the interaction of these genes with the environment. We will never be able to understand them sufficiently well to manipulate them successfully!" The skeptic is only getting started. How many times, he points out, have researchers claimed to identify genes for human characteristics such as aggressiveness, schizophrenia, homosexuality, novelty-seeking, overeating, even bedwetting, only to have their claims questioned or dismissed by other researchers who were unable to replicate their results?[2] The biggest gaffe of them all was the belief that there were more than 100,000 genes in the human genome. Most scientists now accept a much lower number, closer to 30,000, after the announcement in 2000 that the human genome had been successfully sequenced.

Not only does this error show that we know less than we think we do, the critic asserts, but the fact that there are far fewer genes means that their actions and interactions must be even more complicated than we imagined, for otherwise how can so few genes account for the multitude of the structures and functions of the human body? Even if we understood how these genes functioned, we will never be able to manipulate them to produce complex effects such as improved intelligence or finer motor coordination. And if we tried, we'd invariably end up with numerous failures, like the 277 sheep embryos that died in order to yield one cloned Dolly.[3] While this failure rate may be acceptable in animal husbandry, the skeptic reminds us, it would never meet the conditions for ethical experimentation in humans. Finally, the skeptic concludes, we are being far too genetically deterministic by focusing so much on the role of genes without emphasizing the role of the environment in affecting human behavior and traits. Numerous studies of genetically identical twins separated at birth and reared by different families have demonstrated that they do not turn out the same. "You are emphasizing 'nature' to the exclusion of 'nurture,'" the skeptic cries. He has even coined a new epithet: "You are being genist!"

Much of what the skeptic says is correct. Researchers have yet to identify many genes associated with non-disease traits. The interaction of these genes with each other and with the environment undoubtedly will prove complex. Sorting these things out and understanding how to manipulate them will take time and effort, and will be marked by numerous wrong turns and failures. There is a chance that we may never be able to produce wholesale changes successfully in the human genome.

But there is a good chance that we will. Remember that we are dealing with two revolutions occurring simultaneously. Consider how far and fast we have come in our knowledge of human genetics, and in the ability of our computers to process data. The table of elements was created by Russian chemist Dmitri Mendeleev in 1868, and some of the elements, like mercury and arsenic, had been known for centuries.[4] In contrast, it has taken less than forty years from Watson and Crick's discovery of the physical structure of the DNA molecule, initiating the modern revolution in genetics, to the sequencing of virtually the entire human genome. And you have only to read your e-mail to realize how quickly and profoundly the computer revolution has unfolded.

Still, at least a portion of the rescue scenario at the beginning of this book will probably remain science fiction. Major changes in the physical structure of the body—like the specially developed climbing fingers of the professional rescuer—would have to be programmed genetically at a very early stage of an individual's development, perhaps even in a test tube before a fertilized egg was implanted in the womb. If this technology were perfected, it is unlikely that parents would waste it on highly selective enhancements like special finger pads for their children. They would be more likely to install abilities that conferred broader advantages, so that their children could excel not just at mountain rescue, but in a wide range of social spheres. In short, if genetic enhancement becomes a reality, parents probably will opt for traits like beauty, a photographic memory, or intelligence.

What does this mean for society? Will these children garner a disproportionate share of societal benefits? Will the gulf between the privileged and the poor widen even more? Can democracy survive in such a polarized society? And what happens if the genetic changes that are installed ultimately are so great that we no longer recognize these children as human?

The skeptic may be right that this may not be possible in the near future. But the implications are so profound that we cannot afford to take a chance. And even if the ability to alter the human species is remote, we'd better begin preparing for it.

2

Scientific Foundations

In the nuclei of the cells in our body are chromosomes. These are long strands of a molecule called DNA (short for deoxyribonucleic acid), wound up tightly to fit in the confined space of the nucleus. Most normal cell nuclei contain two sets of these chromosomes, one set inherited from the mother and one set from the father. Each set normally contains twenty-three chromosomes. Twenty-two of these chromosomes are referred to by number—chromosome 4, chromosome 21, and so on. The remaining chromosome in each set is a sex chromosome, either an "X" or a "Y." A person who has two X chromosomes is female; a person who has an X and a Y is a male.

Human chromosomes can be viewed and photographed through a microscope. The resulting image is called a karyotype (see fig. 1).

Occasionally, people have an extra chromosome in their cell nuclei. This occurs when something goes wrong in the way the cells divide during reproduction. A person who has an extra copy of chromosome 21, for example, has Down syndrome. This is why Down syndrome is also called "Trisomy 21": "trisomy" means "three copies." Some males have an extra Y chromosome. At one time, this was thought to make them prone to violent criminal behavior.[1]

If you took a chromosome, unwound the DNA, and looked through a powerful microscope, you would see the remarkable physical structure that James Watson and Francis Crick discovered in 1953. The DNA resembles a twisted ladder, a shape called a "double-helix." The ladder is composed of a double string of chemicals called "nucleotides," each of which is made up of a sugar, a phosphate, and something called a "base."

To understand the structure of DNA, think of the ladder cut in half down the middle of the rungs. The sugars and phosphates of the nucleotides form the long strip which is the side of the half-ladder. Sticking out from the side are the bases, each forming a half of a rung.

The bases are the important part of the DNA molecule. There are only four bases: adenine, guanine, cytosine, and thymine, known by their initials A, G, C, and T. They can connect or "bond" to one another only according to a strict set of rules. "A" bonds only with "T" and vice versa. "G" bonds only with "C."

In humans and other higher organisms, the two halves of the DNA molecule in the chromosomes spend most of their time joined together, like what you would get if you reunited the two halves of the ladder. Each of the bases that make up a half-rung is connected to the base of the other half-rung according to the bonding rules: an A with a T, a G with a C. Each rung thus consists of a pair of bases, known as a "base pair." The ladder twists around a central axis, forming the famous double helix shape (fig. 2).

If you unwound the DNA in the chromosomes in a human cell nucleus and laid it in a straight line, you would get a double strand more than five feet long. If you added up the base pairs, there would be about three billion. The DNA of lower organisms has fewer base pairs. The DNA of the bacterium *E. coli* has about four million. Yeast has about fifteen million.

The bases are the important part of the DNA molecule because the order in which they occur, or their "sequence," forms the genetic code of the organism. Along certain stretches of DNA, the sequence of base pairs contains instructions for making proteins. Proteins consist of long, complex chains of chemicals called amino

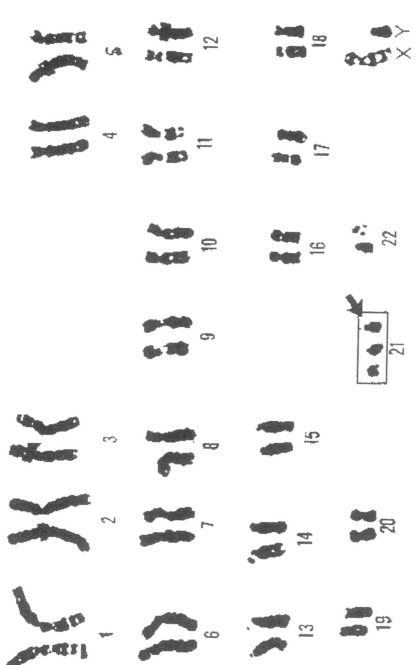

Figure 1. From an actual photograph. *Source:* U.S. Department of Energy Human Genome Program, http://www.ornl.gov/hgmis

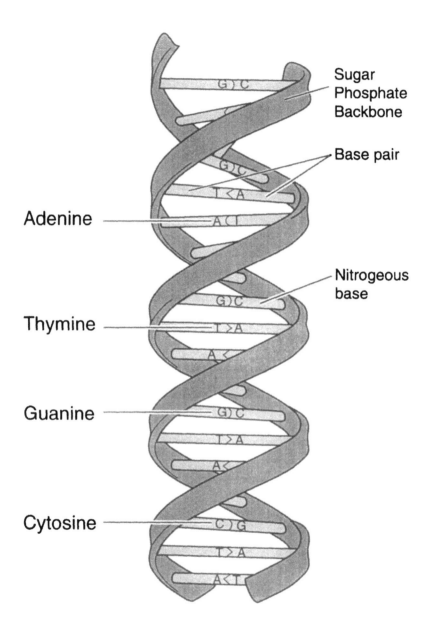

Figure 2. *Source:* National Human Genome Research Institute, Division of Intramural Research, http://www.genome.gov

acids, and they form the structural components of cells and tissue and the chemicals that control biochemical processes, called "enzymes." The stretches of DNA that contain the instructions for making proteins are called genes. (The rest of the DNA was originally thought to have no function, and was referred to as "junk DNA."[2] Scientists are now discovering that the non-gene stretches of DNA do perform some functions, such as containing the instructions for telling genes when to make proteins—in effect when to turn the genes on or off.)

Sometimes the instructions contained in the genes for making proteins are erroneous. For example, all human beings have a gene on chromosome 4 that contains the instructions for making a protein called huntingtin. One stretch of that gene consists of a number of repetitions of the bases C, A, and G. Most people have twenty-six or fewer of these repetitions. But some people have more repeats, and if they have too many, called a "genetic stutter," the gene will not properly produce the protein. This causes the disorder called Huntington disease, named after the physician who first diagnosed it in the nineteenth century.[3] Huntington disease is a neurological disorder that usually doesn't show symptoms until the victim reaches middle age. The most famous victim was the folk singer Woody Guthrie. Sufferers from the disease begin to tremble, and their mental functioning becomes impaired. Eventually they shake so much that they die.

Another example of a genetic disease is cystic fibrosis. Everyone has a certain gene called the CFTR gene that affects the transport of the chemical chloride around the body. In the most common form of cystic fibrosis, three bases that code for a single amino acid are missing. Victims' lungs fill up with mucous and their pancreas malfunctions, interfering with digestion.[4]

The connection between genes and diseases isn't always this clearcut. Some women have certain mutations in their genes that make them susceptible for getting breast cancer, but the mutation seems to be involved in only about 50 percent of inherited forms of breast cancer. Some genes only seem to produce illness when the individual is exposed to some environmental hazard, such as a toxic chemical. Many genes appear to work in tandem, so that diseases only occur when multiple genes malfunction.[5]

Most DNA sequences are the same from one person to another. But about one-tenth of one percent are different. These differences account for why people have different physical characteristics, such as eye color, and why some people get genetic diseases. The differences in DNA also allow scientists to tell people apart by comparing small amounts of their DNA. This is the basis for DNA identification used, for example, in criminal cases.

None of this was known 150 years ago. That was when a Czech monk named Gregor Mendel observed that pea plants appeared to inherit some sort of "factor" from parent plants that explained what color the pea plant would be. Mendel deduced that each plant inherited one factor from each parent and passed on one factor to each of its offspring plants. Mendel noted that offspring sometimes exhibited color traits of neither parent plant, and he figured out how the factors interacted to produce different traits in different pea plant generations.[6]

Some thirty-five years later, scientists observing how cells divided to make sperm and eggs saw certain objects that lined up in pairs and then divided, with one set of the objects ending up in each of the eggs or sperm. The scientists stained the objects so they could see them better, and they named them "chromosomes," from the Latin for "colored things." In 1909, a Danish botanist coined the term "gene" to describe Mendel's "factors."[7] Researchers like Alfred Sturtevant began to mate fruit flies to determine how genes controlled certain physical traits and to chart the locations of the genes relative to one another. Other researchers conducted these investigations by mating corn and other plants and animals.

A big breakthrough occurred in 1953 when Watson and Crick discerned the structure of the DNA molecule and the rules that governed the bonding of A, G, C, and T, the four nucleotides. These discoveries enabled biologists to understand how DNA encoded genetic information and how it transmitted the information when it divided. In the 1970s, recombinant DNA techniques were developed that allowed scientists to reproduce segments of DNA. This permitted them to begin to decipher the order of DNA segments on the chromosomes.

The stage was now set for the Human Genome Project.[8] In the mid-1980s, several U.S. scientists independently conceived the idea of a large-scale project to sequence the human genome. One of them was Charles DeLisi at the Department of Energy, who began to use departmental discretionary funds to get the project started. Ironically, the National Institutes of Health only began to seek federal funding for genome sequencing after prompting by members of Congress. In 1988, Congress appropriated $13 million to the two agencies. By then, small programs were also underway in several European countries. In 1989, the NIH established the National Center for Human Genome Research, with James Watson at its head. Most people count 1990, when the NIH center first gained the ability to make direct research grants, as the actual start of the U.S. government's Human Genome Project.

The ultimate goal of the Human Genome Project was to sequence the entire human genome. The first step was to "map" the genome—to identify the regions in the chromosomes where specific genes were located. This was accomplished with the use of genetic "markers," stretches of nucleotide bases that were known to lie near specific genes, possibly including part or even all of the gene. The proximity of a marker to a gene was determined by studying family inheritance patterns for genetic diseases. If people with Huntington disease, for example, typically displayed a unique sequence of nucleotides along a certain stretch of a particular chromosome, the chances were that this nucleotide stretch was a marker for the Huntington disease gene, and that the gene itself was located near or within it.

The next step was to sequence the nucleotides in proximity to the markers and determine which stretches of DNA were genes—that is, coded for amino acids. When the sequenced fragments were arranged in order on the chromosomes, the result was a sequence map of the genome, what President Clinton in June 2000 called "the most wondrous map ever produced by humankind."

3

Four Revolutions

The unraveling of the mystery of the genome has led to a number of developments with profound implications for society. To understand the prospects for genetic enhancement, it is first necessary to understand these developments.

THE REVOLUTION IN FORENSIC GENETICS

One of the first applications of new genetic knowledge to have a major impact on society was the use of DNA to identify people. This has given rise to the new field of *forensic genetics*. While this is the most mature and well-accepted societal use of genetic knowledge, many aspects remain controversial.

As noted in Chapter 2, the precise sequence of nucleotides in human DNA varies slightly from one individual to another. Some of these variations occur in the genes, the stretches of DNA that code for proteins, and sometimes this causes or makes people susceptible to diseases, like Huntington disease or breast cancer. Nucleotide differences also occur in the DNA regions that do not code for proteins, and therefore do not directly affect a person's health or the way the person looks or behaves. In some of these segments of non-coding DNA, nucleotide differences—such as different numbers of repeating nucleotides—occur fairly frequently, so that if you took DNA from a number of individuals and sequenced the DNA in these segments, you would be likely to find small differences between the individuals. Consequently, if you took two samples of DNA, compared the sequences in a number of these DNA segments and found that both samples had the same sequences, the chances would be high that they came from the same person. This is the basis for modern forensic genetics.

Forensic genetics first compared individuals on the basis of blood types and serum proteins, which follow family patterns of inheritance, rather than on the basis of DNA sequences. These techniques were occasionally used in criminal cases, but more often they were used in civil cases like paternity disputes, where the standard of evidence is less stringent. Then, in 1985, Alec Jeffreys and colleagues in Britain used a comparison of DNA segments to exonerate one suspect in two rape-murder cases and to convict another.[1] In 1987, DNA evidence was used for the first time in the United States to obtain a conviction in a rape case in Florida.[2] Judges struggled to understand the new evidence and to decide if it was admissible and how much weight it should be given. Gradually, they came to accept its value in identifying suspects, in helping to establish guilt, and in exculpating the innocent.

The DNA used in forensic genetics can come from a variety of sources: blood, semen, hair, saliva, even cells scraped from the skin or from the inside of the cheek. DNA found on a crime victim or at a crime scene can be compared to DNA from a suspect to see if they match. An important technical breakthrough was polymerase chain reaction, which allows minute amounts of DNA to be copied many times, or "amplified," so that small samples of DNA can be analyzed and compared. A current technique uses "short tandem repeats" (STRs) that comprise only three to five nucleotide base pairs; the FBI has identified thirteen STRs at different places on the human genome as the standard battery for identification tests.

In early DNA testing, a smaller number of DNA segments were compared. As a result, the probability that two matching samples came from different people was fairly high. In *New York* v. *Castro,* a murder case in 1989, for example, the testing laboratory reported that the probability was on the order of one in a million that the DNA from a watch found at the crime scene came from someone other than the suspect.[3] At first this may seem like overwhelming evidence. But there might be as many as 250 Americans whose DNA could have been deposited on the watch, and who therefore could also have committed the crime. For the evidence to be convincing beyond a reasonable doubt, which is the standard for a criminal conviction, there would have to be additional evidence linking the suspect to the crime, like living or being seen near the scene, or a motive. Judges had to be on guard for what was known as the "prosecutor's fallacy"—attempting to establish guilt on the basis of what sounded like an impressive statistic despite the possibility that the DNA came from someone other than the suspect.

Advances in DNA testing have dramatically increased the statistical power of the evidence. If two samples of DNA match at all thirteen of the STR locations in the FBI standard battery of tests, the chances that they come from different people is extremely remote: for Caucasians in the U.S., about 1 in 575 trillion.[4]

This statistical power gives DNA evidence a persuasive role in criminal cases. But care still must be taken lest the wrong person is convicted and, for a capital offense, perhaps put to death. Although laboratories take elaborate security precautions, they can still mix up or contaminate samples. In the *Castro* case, for example, the laboratory so botched its technique that it ultimately turned out to be impossible to say if the samples of DNA from the watch and from the defendant matched at all. Even if laboratory practices are followed properly and there is no question that the suspect's DNA was found on the victim or near the crime scene, this still does not necessarily mean the suspect is guilty of a crime. A match from semen can indicate rape, but it also can result from consensual sex; a suspect whose blood is found on a murder victim may have acted in self-defense.

DNA evidence not only can help prove guilt; it can help establish innocence. The FBI reports that since 1989, DNA tests have exonerated approximately 25 percent of sexual assault suspects whose DNA was analyzed.[5] In the ten years since it was founded at Cardozo Law School, a program called the Innocence Project has freed about sixty-five prisoners, eight of whom were on death row.[6]

DNA evidence also can be used in legal disputes other than criminal cases. It can establish biological parenthood by showing that the DNA of a child could not have come from the father or, in less frequent cases, from the mother. All courts now accept DNA evidence in paternity cases. DNA also can be used to identify human remains. During the Gulf War, no U.S. personnel were buried without having been identified.[7] DNA testing is being used to identify remains from the World Trade Center attack. And the government is seeking DNA samples from relatives of Osama bin Laden to determine if his remains lie currently unidentified somewhere in Afghanistan.[8]

The increasing usefulness of DNA for forensic purposes has led to the creation of DNA banks. The U.S. military collects and stores blood samples from all enlisted personnel and conducts and stores the results of DNA tests on the samples. The bank can hold eighteen million blood samples; it currently stores about three million.[9] States have created DNA data banks to store the results of tests conducted on DNA samples collected from convicted criminals. Law enforcement officials increasingly are using the data to make "cold hits"—identifying suspects solely on the basis of a computerized match with the results of DNA testing stored within the bank. The FBI operates the Combined DNA Identification System (CODIS) under the authority of the DNA Identification Act of 1994.[10] In addition to DNA profiles collected by the FBI itself, CODIS stores information from state data banks and maintains it in two databases: the Convicted Offender Index and a Forensic Index that contains unidentified DNA test results from crime scenes. The FBI hopes to use the Forensic Index to link crimes, ultimately leading to the identification of the perpetrators.[11]

While DNA data banks have many valuable uses, they raise a number of concerns. How comprehensive should they be? Initially, states collected DNA only from sex offenders.[12] Should states and the FBI expand the database to include DNA test results from all convicted criminals? Or just from felons, or from those who commit violent crimes? Twenty-three states collect and test DNA from persons convicted of misdemeanors.[13] Delaware's program includes all misdemeanors against children, including selling them tobacco products and giving them tattoos. Currently, three states—Louisiana, Mississippi, and Kentucky—collect DNA from all persons arrested for sexual felonies, not just from those who are convicted. Should DNA be collected from everyone who is arrested?

How long should the information be retained? What if a conviction is later overturned? When someone is arrested, they typically are fingerprinted. The prints are stored even if the person is later exonerated. (The finger- and footprints that hospitals obtain at birth are not routinely stored.) Should the same be the case with DNA information? Illinois law prohibits DNA records from being expunged. Only five states require expungement if the person ultimately is found innocent. What about juvenile offenders? Twenty-six states obtain DNA from juveniles. Their criminal records are expunged when they reach the age of majority, but none of these states requires the DNA portion of their record to be expunged or sealed, and one court in Arizona has upheld this practice in the face of a constitutional challenge.[14]

In the wake of the World Trade Center attacks, pressure is mounting to expand the state and FBI DNA data banks. There are calls for collecting DNA from all suspected terrorists who enter the country.[15] Even before the terrorist attacks, Rudolph Giuliani, mayor of New York at the time, said he had no objection to collecting DNA from everyone at birth.[16] It may come as a surprise that, in one sense, this is already being done: State law requires a blood sample to be obtained from all newborns so that it can be screened for certain diseases and conditions. Currently the results of these tests are not stored for DNA identification purposes, but they could be.

DNA databases facilitate cold hits. Although they make crime solving and the identification of remains easier, there are also risks. There can be mistaken matches. The larger the database, the greater the likelihood that an apparent match is an error and the greater the chance that the person suspected as a result of the match did not commit the offense. With the chance of a random match being in the range of 1 in 575 trillion, it seems reasonable for law enforcement to use cold hits as a first step in solving a crime. The more difficult issue is whether convictions should be obtained solely on the basis of such DNA evidence.

Instead of establishing comprehensive databases, the police could conduct "DNA dragnets," in which DNA is collected from all people in the vicinity of the crime, or all people fitting a certain description. In fact, the first use of DNA evidence to solve crimes—the 1985 rape-murder cases in Britain—involved a DNA dragnet in which DNA was obtained from blood and saliva samples from more than 5,000 people—all males between the ages of seventeen and thirty-four in several villages near the crime scene.[17] But this creates the risk of stigmatization, particularly if the suspect is thought to be from a particular racial or ethnic group. In *The Body Bazaar: The Market for Human Tissue in the Biotechnology Age,* Lori Andrews and Dorothy Nelkin recount the ordeal of Blair Shelton, an African American in Ann Arbor, Michigan, who lost his job and was subjected to numerous checks by police who were looking for a rape suspect that fit the description of a "black male 25–35 years of age and between 5'7" and 6'2" in height." In addition to forcing Shelton to give a blood sample for DNA testing merely because he fit the description, the police stopped more than 700 black men and took blood samples from 170 of them.[18]

Even greater risks arise from DNA tissue banking. The FBI CODIS system is a "data bank." It does not store DNA, but rather DNA profiles—the results of tests conducted on the DNA segments. But the military system is a "DNA bank." It stores actual DNA, in the form of blood samples. So do the states—the so-called "Guthrie cards" containing blood samples from newborns. And the NIH maintains a number of DNA, blood, and tissue banks.

The significance of the distinction between a data and a tissue bank is that, as mentioned earlier, the DNA that is used for identification purposes supposedly comes from regions of the genome that do not code for proteins and that therefore do not reveal anything about the individual's "phenotype"—the person's physical or mental characteristics. But a bank that stores the DNA itself can test that DNA for phenotypic information. One NIH tissue bank, for example, contains DNA from Ashkenazic Jews, obtained as part of a study of Tay-Sachs disease, a genetic disease that is prevalent in that population. Researchers studying the DNA subsequently discovered a correlation between certain genes and the risk of breast and ovarian cancers, even though the DNA was not collected for that purpose. As the number of genes known to code for diseases, conditions, and other characteristics grows, and additional genetic tests for these traits are developed, these tests can be performed on the DNA stored in tissue banks. One problem this raises is whether DNA stored with permission to use it for one purpose—researching Tay-Sachs, for example—can be studied for another purpose without the consent of the people from whom the DNA was obtained. Another concern is that, without their knowledge or consent, the test results could be released to employers, insurers, family members, the press, or the general public, and could be used to discriminate against the individuals whose DNA has been stored.

Even if only DNA test results, and not the DNA itself, is stored, creating large DNA databases and conducting DNA dragnets raises the specter of George Orwell's *1984* and "Big Brother," with the government requiring everyone to carry DNA identification cards and to submit to frequent DNA testing to solve crimes, to track illegal aliens, to gain entry to sensitive areas, maybe even to confirm that they paid their taxes. The war on terrorism is likely to accelerate this process by increasing the interest in having a highly accurate means of identifying people. DNA testing may become commonplace. Indeed, as we will see, creating and maintaining massive, population-wide DNA databases may be necessary in order to cope with the challenges of genetic enhancement.

THE REVOLUTION IN GENETIC INFORMATION

A second revolution in human genetics is the *revolution in genetic information.*

In contrast to forensic genetics, in which segments of DNA from different sources are compared to see if they come from the same person, the revolution in genetic information involves testing DNA to obtain knowledge about the characteristics of specific individuals or groups.

Until recently, knowledge about an individual's genetic makeup was based on indirect information. This might be a person's physical appearance, such as eye or hair color. It could be symptoms of genetic illness, like the trembling or "chorea" exhibited by persons with Huntington disease, or it could be chemical evidence of genetic malfunctioning, like elevated chloride levels in the sweat of people with cystic fibrosis.

One of the most important indirect sources of genetic information is a person's family history, the description of the illnesses and causes of death of the members of that person's immediate family. These conditions may be heritable, and if so, the family history can predict a great deal about an individual's own susceptibilities to disease. A person whose parents both died of heart disease at an early age, for example, may have inherited genetic factors that predispose that person to the same fate.

Genetic information also can be gleaned by examining chromosomes. A karyotype, or visual image of a person's chromosomes, can be made by staining cell nuclei. This allows the detection of chromosomal abnormalities like Down syndrome, caused by an extra copy of chromosome 21. For a number of years, chromosomal analysis has been performed on fetuses using amniocentesis (extracting amniotic fluid with a needle) or chorionic villi sampling (extracting placental fragments with a needle). Another type of prenatal genetic test measures the amount of maternal serum alpha-fetoprotein in a pregnant woman's blood; this can identify neural tube defects such as spina bifida (a condition in which the spinal column does not completely close over the spinal cord) or anencephaly (in which part of the brain is absent). All this information can be used by parents to decide whether or not to bring a fetus to term, and in a few situations, surgery and other treatments can be performed on the fetus to help correct the problem.

The discovery of genetic markers—stretches of nucleotides that are physically located near genes—provides another source of significant genetic information. By identifying distinctive markers within families, geneticists can determine if members of the family are likely to have inherited genes for particular diseases, such as Tay-Sachs, cystic fibrosis, and sickle cell disease. This could be accomplished even before researchers knew exactly where the genes were located or the actual sequence of their DNA.

As the Human Genome Project has unfolded and the actual

genes themselves have been identified, new genetic testing techniques have been developed to analyze the nucleotide sequences themselves. Some aspects of the techniques being employed are the same that are used in DNA identification testing, for example making many copies of short stretches of DNA. The DNA sequencing process increasingly is being automated along the lines of the process Celera used to sequence the human genome. One of the more recent advances is the DNA chip, a small glass plate that contains a large number of strands of synthetic DNA. When a sample of a person's DNA that has been labeled with fluorescent dye is placed on the chip, the DNA binds with the synthetic DNA, and when the person's DNA matches the DNA on the chip, it glows a certain color. For example, if the synthetic DNA on the chip includes the huntingtin gene with a normal number of nucleotide repeats, and the color reveals there is no match, this shows that the person has more than the normal number of repeats and may have or in the future may display symptoms of Huntington disease. Since the chip can be loaded with a large number of synthetic DNA sequences, it can simultaneously test for a large number of mutations.

DNA chips integrated into automated, high-speed sequencing machines create the possibility that a sample of DNA can be tested rapidly to reveal a person's entire genetic profile. Sequencing time has dropped dramatically in just a few years. Within the next decade, it is likely that we will be able to sequence your entire genome while you wait.

A bigger challenge will be to make sense of the results. The interaction of genes and of genes with the environment is extraordinarily complex. Numerous different mutations can be associated with a single gene, each affecting the functioning or malfunctioning of the gene in different ways, often with subtle effects. Unraveling the meaning of the streams of data that will pour out of high-speed sequencers will require powerful computer programs designed with a sophisticated understanding of the underlying genetic science. The development of such analytic tools will take time and will occur gradually. But the utility of the results will ensure that development takes place as rapidly as possible.

A separate challenge will be to communicate the results to the individual. Potentially, there will be information about many different susceptibilities and characteristics. Except in the case of so-called single-gene conditions like Huntington disease, where genetic test results provide a relatively straightforward prognosis, high-speed genetic sequencing will yield a blizzard of variations and probabilities. Learned intermediaries, such as well-trained geneticists and genetic counselors, backed up by expert computer programs, will be required to explain the meaning of the results to the affected individuals. The information will contain sobering and perhaps dire predictions about future health status and longevity. People will hear the same information in different ways, and their tolerances for good and bad news will vary. Hence, methods will be needed to convey the information in a compassionate and highly individuated manner.

Despite these hurdles, genetic testing already is changing the practice of medicine in significant ways. In a growing number of cases, diseases can be diagnosed more rapidly and definitively. A person may experience muscle stiffness for a number of reasons, and it may be difficult to determine exactly what is wrong, but now genetic testing can pinpoint if the person is suffering from certain genetic illnesses, such as myotonic dystrophy. Some of the early symptoms of Huntington disease can be confused with other conditions, such as dementia. Again, genetic testing can reveal the telltale presence or absence of excessive numbers of nucleotide repeats on chromosome 4.

Increasingly, genetic testing also can predict a person's risk for future disease. This can facilitate early treatment or prevention. One common genetic disorder that can be detected before symptoms arise is hereditary hemochromatosis. The illness occurs in 1 out of every in 200 Caucasians, and can be readily treated presymptomatically with phlebotomy (a technique similar to the ancient practice of bleeding).

More often, genetic testing does not tell that a person has a genetic illness, but only that the person has a certain risk of having it later on in life. Sometimes people can take preventive measures to reduce this risk, like modifying their diet and exercising more if they have an inherited risk of heart disease. In other cases, preventive steps may create risks or harms of their own. This raises questions about how accurate the predictive testing should be and how great the risk of genetic illness is before people are advised to take preventive steps.

Take genetic testing to detect the risk of breast cancer. In the mid-1990s, researchers studying the DNA data bank that the NIH maintained on Ashkenazic Jews discovered mutations in two genes, BRCA1 and BRCA2, that were associated with an elevated risk of breast cancer in women in that population. Commercial tests were quickly developed. Some women with a positive test result have undergone a procedure called prophylactic radical mastectomy, in which their breasts are surgically removed in an attempt to eliminate the starting point of the disease. The problem is that the predictive value of the genetic tests is limited. Most forms of breast cancer may not be inherited: Only about 5 to 10 percent of the 180,000 women in the United States who are annually diagnosed with breast cancer appear to suffer from an inherited form of the disease. Furthermore, the mutations on the BRCA1 and BRCA2 genes are not completely "penetrant." That's a term geneticists use to describe a situation in which the same mutation produces widely differing effects in different people. While some people with the BRCA1 and BRCA2 mutations will develop virulent forms of breast cancer, other women with the same mutations will not develop the cancer at all. Even in women from families with high rates of breast cancer, between 10 and 15 percent of those with the BRCA1 mutation will not develop the disease. All this makes sound decision-making based on the test results very difficult.

Nevertheless, as more genes are identified and their effects understood, more genetic tests will become available, and they will have greater predictive value. There are now about 500 commercially available genetic tests.[19] The number grows daily. Moreover, as noted earlier, scientists are developing faster and faster ways to decode DNA. It is entirely possible that within the next few decades, geneticists will be able to take a sample of DNA and deliver to the person a complete list of the person's genetic illnesses and susceptibilities.

One question is: Who will have access to this information and what use they will make of it? Individuals have an interest in obtaining their genetic profile. But not everyone may want the information. Some people may not want to know what future ailments they are susceptible to if they can do nothing to prevent or postpone them. Although genetic testing for Huntington disease has been available for a number of years, many members of at-risk families still refuse to be tested. At present, there is no treatment or prevention for the disease. Living with uncertainty, which leaves open the possibility that the individual has not inherited the disease mutation, may be preferable to living with a death sentence.

Given the choice, however, others may feel differently. Some people might want to know if a spouse or potential spouse has inherited a susceptibility for a life-disrupting genetic illness. Adolescents and even children may want to know if they have inherited a disease-causing mutation. Some genetic diseases only manifest themselves if a child inherits a mutated gene from each parent. Potential spouses might want to know if they both carry such a "recessive" disease mutation, in which case their children may be affected even though the parents are not symptomatic. In some Orthodox Jewish communities, in which marriages are still arranged by matchmakers, teenagers are tested for the mutation for Tay-Sachs disease, an untreatable, recessive genetic illness that kills children within a few years of their birth.[20] The teenagers are not told the results of the test, but the head rabbi is. Before a marriage takes place, the matchmaker consults the rabbi, and marriages between two carriers of the Tay-Sachs mutation are forbidden. To preserve privacy, the couple is told simply that they are not a good match—which could be for any number of reasons.

The alternative to this type of testing program is to proceed to conceive a child, then test the fetus and abort it if is affected. But abortion carries with it health risks to the mother, and it offends the morality and religious convictions of some individuals. Even those who believe that abortion is a choice that must be left to the mother without government interference, at least in the early stages of a pregnancy, may be troubled by the notion that parents should be free to test and abort fetuses that have inherited conditions that the parents deem to be an unbearable affliction, no matter how minor. Advocates for the rights of persons with disabilities, for example, object that it is unethical to abort a fetus merely because it has Down syndrome or a hearing deficit. In India and China, parents use ultrasound to determine the sex of a fetus and abort those—namely, females—that are the wrong gender.[21] Even if such gender-based abortions are legal in the United States, they raise vexing questions about discrimination and the limits of reproductive freedom, which we will return to later.

One alternative to testing and aborting a fetus is to employ in vitro fertilization (IVF), in which the egg (either the mother's or a donor's) is fertilized in a test tube with sperm (either the father's or a donor's) and then implanted in the womb (either the genetic mother's or a surrogate's). Typically a number of eggs are fertilized during the IVF process. The resulting embryos can be tested genetically before any are implanted, and only those that are healthy can be introduced into the womb. Lately, there have been reports that some clinics are even testing embryos for late-onset conditions like Huntington disease and Alzheimer's, so that parents can avoid implanting embryos with these susceptibilities.[22]

The question arises: How much discretion to decide which embryos to implant should parents have? For example, should they be able to select gender, perhaps as an alternative to ultrasound followed by abortion? Some infertility clinics that perform IVF offer a technique called "sperm sorting,"[23] in which sperm are segregated centrifugally based on their weight, since sperm cells that contain two X chromosomes, and that therefore will produce a female child, are heavier than sperm cells containing the male combination of an X and a Y chromosome. So far this procedure remains available, although it provokes almost as much criticism as gender selection by abortion. But the ability to test embryos before they are implanted will become important when we consider genetic enhancement.

If a genetic illness is preventable or can be made less severe by early detection and treatment, the privacy interests that lie behind efforts like the Orthodox Jewish testing program for Tay-Sachs become less compelling. Family members may assert that they have a right to know if someone in their family is affected so that they can take preventive steps. There are a growing number of court cases in which physicians have been successfully sued for failing to inform individuals that a family member has an inherited illness that can be treated if caught in time. One woman suffering from a hereditary precancerous condition called familial adenomatous polyposis successfully sued a physician on the ground that he had failed to notify her that she might have inherited the condition from her father.[24] If she had been warned, she pointed out, she could have had her colon removed as a precaution, giving her an excellent chance that she would not get the cancer. What is noteworthy about the case is that the defendant had not treated her, but her father—twenty-seven years earlier. These events happened so long before that the physician had since died (along with the patient's father); the suit was filed against the physician's estate.

Although the growing ability to conduct genetic testing is making it easier to diagnose and treat genetic illnesses, it also raises fears that the test results will be used to discriminate against people who are tested or their families. Health, life, and disability insurers might use information about genetic susceptibilities for future illness to deny insurance or to charge unaffordable rates. Employers, who foot at least part of the bill for most people's health insurance, might use the information to refuse to hire certain job applicants or to fire existing employees.

At present, there is little evidence that insurers or employers are using genetic testing to discriminate against persons who may be at risk for genetic diseases. This may be because there are still only so many susceptibilities that can be tested for, because employers and insurers remain relatively uninformed about the testing that is available, or because they perceive the costs of testing to outweigh the benefits. But it also is extremely difficult to find out what insurers and employers are actually doing. How many of them will freely admit that they discriminate against people based on their inherited characteristics? How often do they tell people the real reason why they don't hire them, or why they won't sell them an insurance policy? How much do job or insurance applicants know about what laboratory tests are performed on blood samples obtained during employee or insurance physicals? And in considering the prevalence of genetic discrimination, remember that genetic information can come from other sources besides genetic testing. A person might fail a physical because they are obese, or have high blood pressure or diabetes, or because of what killed their parents, all of which may relate to their genes.

While it is unclear that much genetic discrimination is taking place, fears that it might occur in the future have prompted Congress and a number of state legislatures to enact laws prohibiting genetic discrimination. The laws form a patchwork of loopholes and protections. Some of them, like Ohio's, are narrowly defined to forbid access only to the results of DNA tests; under Ohio law there is nothing illegal, for example, about denying someone insurance because of a family history of genetic disease.[25] Other laws, like New Jersey's, broadly prohibit insurers from obtaining or acting upon any "genetic information," which is defined to include information obtained by physical examination or a family history. The laws vary from one state to the next, covering different types of genetic information, different forms of insurance, and different employers.

Congressional action has created a similar hodgepodge of loopholes and uncertainties. The Americans with Disabilities Act protects against employment discrimination, but it is not clear that its protections extend to genetic susceptibilities rather than to current illnesses. The federal agency charged with enforcing the act takes the position that it does, but so far the courts have not ruled on the question, and until they do, the interpretation of the statute remains in doubt. Congress passed a law called the Health Insurance Portability and Accountability Act, which contains a prohibition against genetic discrimination in access to health insurance, but it only applies to larger group insurance plans, while the threat of genetic discrimination applies primarily to individual and small group policies. The Americans with Disabilities Act, meanwhile, contains an exemption for health insurance; insurers may refuse to insure people or may charge them extra because of their genetic endowment (or for any other reason) so long as the insurer bases its decision on sound data showing how much it would in fact cost to insure these individuals.

Regardless of their inadequacies, genetic anti-discrimination laws garner wide support. The public seems to feel that, since people do not have control over what genes they inherit, they should not be penalized if they inherit genetic diseases or susceptibilities. But these broad anti-discrimination laws inadvertently may create problems. People with negative genetic test results or a family history relatively free of genetic illnesses cannot submit that information to an insurer and obtain a lower premium. Instead, they have to pay the same premium as people with a higher risk of disease. This may sound fair and just. Since no one does anything to deserve their genetic endowment, no one is entitled to pay less for health insurance or health care simply because of that endowment. But the effect is that the healthy subsidize the less healthy by paying higher premiums. In some cases, this could drive the cost of insurance up to the point that healthier individuals decide not to purchase it.

Another insurance problem raised by genetic anti-discrimination laws is known as "adverse selection." This is because, although individuals are free to test themselves, insurers are precluded from obtaining the information or acting on the results. If, based on positive genetic test results, individuals know that they are at a high risk for genetic illness, they will tend to buy insurance to cover it. On the other hand, individuals with negative test results, indicating a lower risk, will tend not to. This will drive up premiums, since those who are likely to make claims will come to represent a growing percentage of those who are insured. This will lead more people to get tested and to avoid purchasing insurance if the results are negative, which will drive premiums even higher. Eventually the only people who will consider purchasing insurance will be those who know they will soon need it to pay their health care bills. This will make it cheaper simply to pay for whatever care they can afford out of their own pockets, rather than pay the additional costs of insurance, which includes administrative expenses and profit for the insurance companies. In insurance parlance, this is known as "the death spiral." The only way to prevent adverse selection and the death spiral may be to replace the current system of private insurance with a public program in which everyone is required to enroll and to pay the same premiums, regardless of risk. This is similar to how we fund the portion of the Medicare program that covers in-patient hospital care, but it would dramatically change the way most Americans now get health insurance. This is just one of the wide-ranging effects that the revolution in genetic information may have on society.

THE THERAPEUTIC REVOLUTION

Growing knowledge about genes and their relationship to disease is not only leading to an expanding number of genetic tests, but to new treatments and preventive measures based on genetic technology. The first fruits of this therapeutic revolution have been drugs made with recombinant DNA. This involves identifying genes that produce certain proteins and splicing them into, or "recombining" them with, the DNA of an organism such as the bacterium *E. coli*. The bacteria can be copied many times, creating biological factories that produce large amounts of the protein products. This technique has been used to manufacture therapeutic drugs that previously were available only in small quantities. One example is human growth hormone (HGH), which is administered to children with an inherited pituitary deficiency that causes dwarfism. Originally, human growth hormone was only found in human pituitary glands, and had to be collected from cadavers. The supply was extremely limited and was doled out by the National Institutes of Health. Then drug manufacturers discovered how to use recombinant DNA to engineer bacteria that could produce the hormone. The current supply of this synthetic human growth hormone is virtually unlimited. Ironically, while the supply has increased, so has the price to the patient; the drug manufacturers charge for their product, while the NIH handed out its limited stocks of cadaveric extract for free.

New knowledge about human genetics is also leading to the discovery of new drugs themselves. In his book *Life Script,* for example, science writer Nicholas Wade describes how the drug company Human Genome Sciences, using a large database of genetic sequences, developed a drug called KGF-2 that promotes healing.[26] He cites its CEO, William Haseltine, as saying that in thirty years, virtually all new drug discoveries will come from genetic science.[27] Francis Collins notes that more than fifty gene-based drugs have become available for human use.[28]

Genetics is even helping to make currently available drugs more effective and less harmful. The new field of pharmacogenomics employs genetic testing to predict how an individual will respond to existing drug regimens. A gene test called Cytochrome P4502D6, for example, will soon be available to predict how patients metabolize a number of different drugs, leading to more precise prescribing.[29] The days may be numbered in which patients suffering from depression must spend months and even years trying a variety of anti-depressants, each followed by a washout period, before finding the drug that works for them. Another genetic test can identify cancer patients who are especially sensitive to a standard chemotherapy agent, enabling oncologists to use much lower dosages with fewer side effects.[30] Still another test can reveal if patients with congestive heart failure will respond to drug therapy, or if they require a heart transplant.[31]

The next step in the therapeutic revolution made possible by the new genetics is gene therapy. In September 1990, a four-year-old girl from northern Ohio named Ashanti DeSilva received infusions of genetically engineered white blood cells at the NIH in Bethesda, Maryland, in what is generally considered the first successful use of gene therapy. DeSilva suffered from severe combined immune deficiency, an inherited disorder in which a missing enzyme prevented her immune system from functioning properly, leaving her open to severe infections and forcing her to lead a carefully monitored, highly restricted existence. The white blood cells were removed from her own body, and normal copies of the malfunctioning gene were inserted into their nuclei. After receiving four infusions of the corrected genes over four months, she markedly improved.[32]

DeSilva's gene therapy is only partially successful. She continues to receive traditional drug therapy for the disorder, and since the cells with the normal gene do not continue to work indefinitely, she must receive additional infusions every few months. In April of 2000, however, French scientists announced that they had used gene therapy to treat infants with another form of the same illness.[33] This time, the correctly functioning genes were inserted into the babies' bone marrow, where they can replicate, eliminating the need for repeat infusions. The researchers claimed that the babies' immune systems became normal and that they have stayed that way without traditional drug therapy, although two children appear to have developed cancer as a result of the treatment.

These efforts mark the beginning of a new era in medicine in which diseases will be treated by inserting genes into the patients' bodies. Initial trials have targeted diseases caused by malfunctioning genes, inserting correctly functioning genes to produce missing enzymes. Approaches that might be employed in the future include blocking the functioning of existing defective genes or proteins, or inserting "gene factories" that manufacture therapeutic substances in the body. Ultimately, it may be possible to correct the DNA itself by removing errant nucleotide sequences, such as the extra nucleotide repeats that cause Huntington disease, or by adding nucleotides.

Gene therapy faces a number of technical hurdles. First, the properly functioning DNA has to get into the patient. The most common way of accomplishing this is to insert the desired genetic material into a "vector"—usually an organism, such as a virus— and then insert the vector into the patient. The most common viral vectors are adenoviruses—the common cold—and retroviruses (a class which includes HIV, although HIV obviously is not used as a gene therapy vector). Before a viral vector can be used, it has to be stripped of any disease-causing elements.

A second problem with gene therapy is to make sure the vector delivers the genetic material to the correct site in the patient. A gene therapy for an immune deficiency disorder like Ashanti DeSilva's might work if enough of the correctly functioning genes get into the patient's bloodstream. But a treatment for cancer might fail, or be accompanied by unacceptable side effects, if it cannot be delivered only to the cancer cells.

Another problem is that once the therapeutic genes get into the patient and find their way to the correct site, they must function properly. Every cell nucleus basically contains a complete set of a person's genes, but only some of the genes are "turned on" in particular cells; otherwise, the cells in the liver, for example, would manufacture not only liver enzymes, but enzymes that are produced in the gall bladder or the pancreas. Therapeutic genes must be turned on as well. Furthermore, they must turn on correctly, producing the correct amount of a protein, for example, and not causing unwanted effects, such as cancer (uncontrolled cell division) or sterility. Finally, the body discards all of its cells after a certain interval of time, including the cells containing engineered genes. (Even the oldest tissues in our bodies—our bones—are completely replaced about every three years.) But the gene therapy must continue to function for as long as is necessary to achieve the desired therapeutic effect, which may be longer than the lifetime of the cells in which the therapeutic genes reside. One solution—the one used for Ashanti DeSilva—is to reinfuse corrected cells every so often. An alternative is to use a vector that incorporates the genetic material into cells in such a way that the material is replicated and survives indefinitely. (This is one benefit from using retroviruses as vectors.)

As a result of these technical hurdles, gene therapy has had only limited success to date. Some scientists are skeptical that these hurdles will easily be overcome. Others are more confident. One thing that is clear is that the continuing pressure to find new cures and treatments will mean that enormous resources will continue to be devoted to gene therapy research. If these obstacles can be overcome, they will be.

But there are other concerns besides these technical ones. Ashanti DeSilva may have been the first person to be treated at least partially successfully with gene therapy, but she wasn't the first to be treated. That distinction was supposedly earned by three German girls in 1970, who were infused with a virus containing a correctly functioning gene, but whose inherited liver disease failed to improve. Two other patients who suffered from a hereditary blood disorder participated in a gene therapy experiment in 1980, one in Israel and the other in Italy. Unfortunately, their physician, Martin Cline, the chief of hematology/oncology at the UCLA Medical Center, attempted to insert DNA into their bone marrow without waiting for approval from the review body at UCLA that oversees human experimentation. As a result, he lost the right to receive federal research grants, and he lost his department chair at UCLA.[34]

Neither this incident nor the continuing technical difficulties seemed to deter researchers from pushing ahead with their experiments. But something that happened in 1999 did—the death of eighteen-year-old Jesse Gelsinger.[35]

Gelsinger suffered from a genetic disease called OTD (for ornithine transcarbamylase deficiency). OTD affects the ability to break down ammonia, a by-product of protein metabolism. The disease is especially devastating in newborns, who typically slip into a coma within hours of birth and suffer severe brain damage. Most die within six months.

Gelsinger himself had a much milder form of the disease. He was able to metabolize some ammonia and to control his condition with a combination of enzyme pills and a low-protein diet. But by his eighteenth birthday, he was taking thirty-five pills a day, and the dosage would probably need to be increased as time went on. Plus he was eager to help find a cure for the disease, to save the babies. So he enrolled in an experiment at the University of Pennsylvania in which volunteers with mild forms of the illness were given infusions of properly functioning DNA in an adenoviral vector. This was the first phase of human experiments with the vector. The purpose of the experiment was not to treat the disease, but just to see if the genes, once inserted, went ahead and produced the missing enzyme.

Within twenty-four hours, however, Gelsinger was deathly ill, suffering from multiple organ failure. Four days later, he was pronounced dead.

What caused his death is still not completely understood. Most likely his liver was not functioning well enough for him to have been subjected to even the weakened virus used as the vector. The researchers had tested his liver function when he had applied for the study, but they had not redone the test when he actually had received the infusion several months later.

Gelsinger's death raised all sorts of questions about the process used to design and approve gene therapy experiments. Had the informed consent process adequately explained the risks of the study to Gelsinger? Could a youngster of eighteen understand the true risks? Had the University of Pennsylvania erred when it performed the experiment on a relatively healthy patient like Gelsinger, rather than on the more severely afflicted newborns? Both the NIH and the U.S. Food and Drug Administration (FDA) had reviewed and approved the plan for the study before it began. Had they acted too hastily?

After Gelsinger's death, reports began to surface about numerous other health problems suffered by subjects in gene therapy experiments, including several other deaths. Most of these incidents had been reported to the FDA as required by federal law, but they were not made public because the researchers conducting the studies and the FDA regarded the research as trade secrets that, if revealed, could give competitors an unfair advantage. The adverse events also were supposed to be reported to the NIH if the studies, like the one in which Gelsinger took part, were being supported with federal funding. But this reporting hardly ever occurred, since in contrast to the FDA, the NIH made these reports public to help guide researchers designing future studies. After Gelsinger's death, the NIH received a sudden burst of 691 reports of serious side effects; only thirty-nine had been reported to the NIH when they had taken place, as required by NIH regulations.

The FDA's unwillingness to make these adverse event reports public due to concerns about competitive advantages highlights the enormous financial stakes that are riding on gene therapy experiments. The goal of discovering treatments for dreaded diseases is compelling. Some disorders, like the disease Jesse Gelsinger suffered from, are rare. Others like cancer are just as deadly, and strike millions. Still others like Alzheimer's disease kill slowly, but in terrible ways. Whoever discovers cures for these scourges will reap tremendous economic rewards.

It may come as no surprise, therefore, that questions about the role of economic motivations were raised in the aftermath of Gelsinger's death. Had the researchers and their institutions remained sufficiently free of economic self-interest to conduct ethical human experiments? The study in which Gelsinger died was part of a large-scale gene therapy research program at the University of Pennsylvania's Institute of Gene Therapy, headed by James M. Wilson. At the time, the institute had 250 employees and an annual budget of $250 million. Wilson was also the founder and a stockholder of a company which had the exclusive rights to any of the institute's discoveries. The university was also a stockholder even though, under government regulations, the university's own Institutional Review Board bore the primary responsibility for determining if the human experiments were ethical and safe.[36]

The Gelsinger tragedy also called into question the ability of the federal government to oversee risky genetic experiments. For years, the FDA had maintained that genetic experiments—whether in plants, animals, or humans—were no different from any other experiments aimed at developing new foods or therapies. According to the FDA, they needed to meet no special regulatory requirements. Nevertheless, when recombinant DNA techniques were first developed in the 1970s, the NIH established a special committee called the Recombinant DNA Advisory Committee, or "the RAC," and charged it with the responsibility of reviewing proposals for human gene experiments. In 1997, however, the functions of the committee were transferred to the FDA. Was the FDA up to the task? Gelsinger's death suggested that it was not. Stung by the criticism, the FDA reacted by proposing to adopt regulations under which it would make public reports of serious side effects from two types of experiments: human gene therapy experiments and "xenotransplantation"—transplanting animal organs into humans. This marked the first time that the FDA admitted that gene therapy might raise special regulatory concerns. But it did not dispel the question of whether the federal government could cope with the difficult regulatory issues that would be posed by more exotic forms of genetic manipulation than what killed eighteen-year-old Jesse.

One of these exotic technologies is germ line gene therapy. The experiments on Ashanti DeSilva and Jesse Gelsinger were designed to deliver corrected genes to their livers and bloodstreams respectively. The goal was for the corrected genes to produce enough of their missing enzymes to cure their enzyme deficiencies. But DeSilva's children and the children of infants with Gelsinger's disease might still be born with the same illnesses, and require their own infusions of corrected genes. (Actually Gelsinger did not inherit his illness and therefore could not have passed it on to his children; his enzyme deficiency stemmed from a random mutation in his genes during fetal development.) In the words of famed biologist Lewis Thomas, despite their ability to dramatically improve health and save lives, the therapies tried on DeSilva and Gelsinger, even if they were successful, would only be "half-way technologies."[37]

A bolder possibility looms. If corrected genes could be introduced at an early enough stage of human development, they not only might alleviate the disease in the individual being treated, but might be incorporated into that individual's reproductive cells—their eggs or sperm. As a consequence, when that individual produced children, the children would inherit the corrected genes and be spared the chance that they too would be affected. A more advanced approach would be to replace the errant genes, rather than supplement them, with the corrected ones; in the case of Huntington disease, the DNA with the extra number of CAG nucleotide repeats that causes the disease would be removed and replaced with genes that contained the normal number of nucleotides.

Treatments that change the genetic makeup of reproductive cells so that the alteration is inherited are called germ line gene therapies, in contrast to the non-inheritable type of treatment used in the DeSilva and Gelsinger experiments, which is called somatic cell therapy. There already has been some success with germ line gene therapy in animals, notably in mice with pituitary dwarfism and with Lesch-Nyhan disease. Since the genetic changes have been incorporated into the genomes of these animals, the animals are known as "transgenic."

The prospect of germ line gene therapy in humans has evoked widespread alarm. In September 2001, a group of prominent scientists, lawyers, and bioethicists convened in Boston to call for an international treaty to ban germ line gene therapy, as well as human cloning for reproductive purposes. The prestigious American Academy for the Advancement of Science issued a report in 2000 that concluded, "any somatic genetic therapy applications where there is a reasonably foreseeable possibility of [human inheritable genetic modifications] should not proceed at this time."[38]

Opponents of germ line engineering cite the danger of adverse effects on the offspring, the risk of unforeseeable side effects that only become apparent in future generations, and potential catastrophic effects on the health of the human gene pool. They analogize germ line engineering to reproductive cloning, and they point to the failed cloning attempts that were necessary to produce Dolly the sheep and to reports that Dolly's youthful cells appear to bear features that are characteristic of much older cells, those of her mother.

In light of these concerns, it may come as a surprise that inadvertent germ line manipulations already have taken place. This has occurred as a result of a type of infertility treatment called "ooplasmic transfer," which was first reported to have been used successfully in 1997.[39] This treatment is used in women whose eggs do not properly attach and grow in the womb as a result of deficiencies in the portion of the egg that surrounds the nucleus, called the cytoplasm. The treatment consists of removing the nucleus of the egg and inserting it into an egg obtained from a donor which also has had its nucleus removed and which has healthy cytoplasm. The egg is then fertilized in the laboratory and implanted in the womb of the intended mother. The germ line is modified because a very small amount of DNA is found outside the cell nucleus, in structures in the cytoplasm called mitochondria. When a cell nucleus is removed, the mitochondrial DNA remains. Children born through ooplasmic transfer therefore have cells containing nuclear DNA from their mother and mitochondrial DNA from the donor of the enucleated egg. In short, the children are transgenic.

In 1998, two eminent geneticists, French Anderson and Emil Zanjani, proposed an experiment in which viral vectors bearing corrected genes would be inserted into fetuses affected with the same autoimmune disorder that afflicted Ashanti DeSilva. The idea was that, by intervening at an early stage of development, the corrected genes would be incorporated in the child's natural DNA and perpetuated by means of normal cell division. The researchers acknowledged, however, that since some of the fetal cells might go on to become reproductive cells, the corrected DNA might alter the germ line.

Inadvertent germ line manipulations like these raise difficult questions. Some argue that it may be ethical to alter the germ line inadvertently if that is the best way to cure genetic diseases. Others disagree, insisting that the risks outweigh the potential benefits.

But all agree that there is an important difference between inadvertent germ line changes and the deliberate alteration of the germ line.

Again, some geneticists assert that intentional germ line manipulation may be the only way to completely prevent a genetic illness. Others offer the alternative of in vitro fertilization in which a number of embryos can be created in the laboratory and tested for genetic illnesses, and only those that are disease-free selected for implantation in the mother's womb. But pro-life advocates object to the fate of the remaining embryos, which are either destroyed or consigned indefinitely to a frozen limbo. And geneticists point out that there are some genetic diseases in which, since both parents may have the defective genes, all of the embryos will be affected, leaving as the parents' only preventive measure the use of donor eggs or sperm, in which case the child will not be related genetically to one or the other parent, and will be a carrier for the mutation and thus able to produce affected children of its own.

The drive to create transgenic animals for use as food or as sources of transplantable organs for humans suggests that the techniques of germ line engineering will continue to be perfected in animals. Once they become routinely successful in animals, the goal of completely preventing genetic illness rather than being satisfied with half-way technologies may make researchers willing to try them in humans. As we will see, this may lead to attempts to modify the germ line to alter non-disease characteristics as well.

THE REVOLUTION IN BEHAVIORAL GENETICS

Scientists have long sought to explain human behavior in terms of inherited factors. Traits ranging from homosexuality to perfect pitch have been attributed to genes. Particular interest has focused on traits that lead to anti-social behaviors, like committing crimes. If criminal tendencies were due, at least in part, to a person's genetic endowment, then genetic testing might identify potential criminals before they caused trouble, treatments might be devised to correct the protein errors that were responsible, and perhaps germ line engineering could be employed to eliminate the offending genes from the gene pool.

The search for an inherited basis for criminal and other socially undesirable behavior has led from phrenology—the study of bumps on the skull that supposedly predicted criminal or violent behavior—to reports in the mid-1960s that an unusually high percentage of males institutionalized with violent or criminal tendencies possessed an extra Y chromosome. Most of these claims have been discredited. For example, it turned out that the reason that a higher-than-expected percentage of males with the XYY chromosomes were institutionalized for criminal tendencies was not that the XYY aberration caused criminal behavior, but that it caused impaired mental functioning, and mental impairment is more common in institutionalized criminals than in the general population.[40] Other purported findings with implications for behavioral genetics, such as claims that genes have been identified for reading disabilities, schizophrenia, psychosis, manic depression, and alcoholism, have been discounted after researchers were unable to replicate the results, suggesting that the original association was specious or due to chance alone.[41] Identifying genes associated with behavioral traits is difficult because the traits are likely to be complex, meaning they are caused by the interaction of multiple genes and between genes and the environment, and because the traits themselves are not well-defined and cannot be identified consistently by different observers.

But the search continues. "Mounting evidence from animal and human studies shows that genetics has a role in human behavior," writes Charles Mann in *Science*.[42]

One stimulus for the hunt is the criminal defense bar, which is interested in raising "genetic defenses" on behalf of its clients in order to have them found not guilty or to reduce their punishment. The first genetic defense was asserted in the early 1970s, based on the XYY chromosomal abnormality.[43] The courts so far have rejected the XYY defense, but have indicated their general willingness to accept genetic defenses if the genetic condition meets various standards, such as that it "interferes substantially with the defendant's cognitive capacity or with his ability to understand or appreciate the basic moral code of his society."[44]

Occasionally the defense works. In one bizarre decision, one woman who murdered her son and tried to kill her daughter was declared not guilty by reason of insanity when she began to experience symptoms of Huntington disease, even though the symptoms did not manifest themselves until seven years after the crime.[45] Unfortunately, this case illustrates one of the pitfalls in this area: Judges and jurors typically have a poor understanding of genetic science and might be led to accept a questionable defense. Yet lawyers persist in seeking genetic precursors for criminal behavior. Currently they are exploring a defense based on a genetic mutation known as MAOA deficiency, which has been reported to be associated with impulsive aggression, arson, attempted rape, and exhibitionism.[46]

Apart from being asserted as a defense to culpability, associations between genes and undesirable behavior could make it possible to take preventive measures. For example, criminals could be tested for the offending genes and offered treatment to reduce their anti-social tendencies. Convicted criminals might be eager to volunteer, especially if treatment led to a reduction in their sentence or to early parole. An analogy is the use of a synthetic hormone called Depo-Provera in sex offenders. The hormone decreases testosterone, reducing the sex drive and the inability to control sexual fantasies. Four states, including California, have enacted laws authorizing the use of this form of "chemical castration" as a condition of parole.[47]

But if geneticists discovered genes that actually caused people to engage in anti-social behavior, particularly violent crimes or sexual crimes against children, lawmakers might go even further. They might screen the population to detect potential offenders and treat them long before they committed crimes. This even could be incorporated into newborn screening programs. Or, we might test the children of convicted criminals and treat those who tested positive. If this proved too expensive or administratively difficult, legislatures simply might order that individuals with these heritable genes be prevented from having children in the first place.

Ironically, we have been down this road once before. In fact, the real beginning of the genetic revolution was not the use of DNA evidence in the courtroom in the early 1980s, but the eugenics movement of the period from 1870 to 1950.

Modern geneticists are understandably reluctant to associate themselves with this earlier foray into genetics and social engineering. As Allen Buchanan and his colleagues write in their book *From Chance to Choice: Genetics and Justice,* "the history of eugenics is not a proud one. It is largely remembered for its shoddy science, the blatant race and class biases of many of its leading advocates, and its cruel program of segregation and, later, sterilization of hundreds of thousands of vulnerable people who were judged to have substandard genes. Even worse, eugenics, in the form of 'racial hygiene,' formed part of the core of Nazi doctrine."[48]

What many people don't realize, however, is that eugenics was not the brainchild of Hitler and his cronies. The idea originated in Victorian England, where a man named Francis Galton coined the term "eugenics" and helped found a Eugenics Record Office at the University of London to "study agencies under social control that may improve or impair the racial qualities of future generations either physically or mentally."[49] Interest quickly spread across the Atlantic, where the movement received substantial financial support from leading citizens, including the Harriman, Carnegie, and Rockefeller families.[50] In 1905, a biology professor named Charles Davenport, who would become a member of the prestigious National Academy of Sciences, established a eugenics research facility at Cold Spring Harbor on Long Island, and recruited a high school teacher named Harry Laughlin to direct its own Eugenics Record Office. (Cold Spring Harbor is now a major genetics research facility headed by James Watson.) Laughlin was instrumental in getting Congress to restrict the immigration of certain ethnic groups on the ground that they were of inferior racial stock. He also led a revival in state legislatures of interest in enacting laws that permitted involuntary sterilization of criminals and the mentally unfit.

The zenith of the eugenics movement in the United States was the 1927 Supreme Court case of *Buck* v. *Bell*,[51] in which the Court, with only one justice dissenting, upheld the constitutionality of a Virginia law authorizing the state to sterilize inmates at the State Colony for Epileptics and Feeble Minded. The case was a setup; it was brought to dispel any doubts about the legality of state eugenics programs. Virtually no defense was offered for the putative plaintiff, Carrie Buck, whom the colony proposed to sterilize because she was believed to be "feebleminded," displayed a record of "immorality, prostitution, and untruthfulness," had given birth to a feebleminded illegitimate child, and had been the daughter of a feebleminded woman who also was incarcerated at the colony.[52] As legal historian Paul A. Lombardo points out, these charges were unfounded. Buck had been a normal child in school. Her alleged immorality and prostitution were based on her illegitimate child, who had been born following her rape by the nephew of her foster family, and the daughter, who died at the age of eight from an intestinal illness, far from being "feebleminded" herself, would go on to do fine in school, even earning a spot on the "honor roll." The Supreme Court nevertheless approved the order for Buck's involuntary sterilization. Its opinion, written by Chief Justice Oliver Wendell Holmes for the majority, culminated in the now-infamous exhortation: "Three generations of imbeciles are enough."

The case achieved its purpose. By 1931, a total of twenty-eight states had enacted compulsory sterilization laws, with more than 3,000 sterilizations a year performed prior to World War II. Again, support for compulsory sterilization was widespread. The case was hailed by such progressive reformers as Clarence Darrow, Helen Keller, and Margaret Sanger, who founded Planned Parenthood as a pro-eugenics organization.

Far from being the brainchild of the Nazis, Hitler's eugenics program, which included compulsory sterilization, was heavily influenced by the eugenics movement in the United States. Hitler had read about the Virginia sterilization law at issue in *Buck* v. *Bell* when he was writing *Mein Kampf* in prison in 1924. The German sterilization law, the first law that Hitler enacted when he came to power in 1933, was modeled on a prototype drafted by Harry Laughlin at Cold Spring Harbor. Laughlin was held in such high regard by the Nazis that they gave him an honorary degree from the University of Heidelberg in 1934.

Nor did eugenics-based sterilization disappear after World War II, although it markedly declined. In 1958, 574 sterilizations took place in Georgia, North Carolina, and Virginia, 76 percent of the national total. At one South Carolina facility, 104 inmates, 102 of whom were black, were sterilized between 1949 and 1960. As late as 1974, a federal district court in Washington, D.C., struck down regulations of the Department of Health, Education, and Welfare (now the Department of Health and Human Services) that provided for sterilizations under federally funded welfare programs, after reports that a number of children had been sterilized without the consent of their parents.[53]

Against this grim historical background, modern interest in

behavioral genetics understandably has sparked controversy. In 1993, a psychologist named David Wasserman at the University of Maryland obtained a $78,000 grant from NIH to hold a conference on genetics and criminal behavior. The plan drew protests from individuals who claimed it was a thinly veiled effort to attribute innate criminal tendencies to certain racial groups, particularly African Americans. One opponent, the director of the Center for Psychiatry and Psychology at Wasserman's own university, objected on the grounds that "behavioral genetics is the same old stuff in new clothes. It's another way for a violent, racist society to say people's problems are their own fault, because they carry 'bad' genes."[54] In response, Bernadine Healey, then director of the NIH, took the unprecedented step of withdrawing the grant at the last minute, causing the conference to be cancelled. This led to the belief in academic circles that behavioral genetics, at least the focus on the association between genes and crime, was taboo. But after Healey resigned, Wasserman revised his proposal to make it broader and more critical of behavioral genetics, received a larger grant from the NIH, and finally held the conference in 1995. And since then, there has been a flurry of books and articles exploring the new field.

The eugenics movement that led to the compulsory sterilization laws and that culminated in the Holocaust was concerned largely with preventing the birth of offspring with undesirable characteristics. This is known as "negative eugenics." But there was another side to the eugenics movement, "positive eugenics," which aimed to encourage the birth of children with desirable traits. This too became part of the eugenics program in the United States between the two world wars. State fairs hosted "fitter families exhibits" and "perfect baby competitions," in which people showed off their "prized" offspring right next to stalls displaying blue ribbon cattle and sheep. Schools gave out "goodly heritage" medals to students. The American Eugenics Society sponsored awards for church sermons extolling marriages by "the best with the best." The Nazis jumped on this bandwagon as well, with their "Lebensborn" program, in which public health officials selected Aryan women to breed with SS soldiers and raised the children in specially selected foster families. Meanwhile, in the United States, the U.S. Army Air Corps, with funding from wealthy conservatives, set up something called the "Pioneer Fund," which gave cash awards to pilots and crew members who had more than three children.[55]

So is the eugenics movement a thing of the past? There is clearly widespread interest in positive eugenics in the United States today, although few would know to call it by that name. People sometimes go to noteworthy lengths to produce the best children. They advertise on the Internet seeking to buy eggs or sperm from beauty queens and Nobel laureates. In 1979, Robert Graham set up the Repository for Germinal Choice for the purpose of collecting and distributing sperm from highly intelligent individuals who had made contributions in science and the arts, and from bright, star athletes.[56] In 1984, Graham reported that fifteen children had been born to women chosen as recipients.

None of these positive eugenics activities is directly supported by the government (although Graham's sperm bank might have received tax breaks as a tax-exempt organization). There may seem to be little of the state-sponsored eugenics that characterized the darkest aspects of negative eugenics. But one branch of the government—the judiciary—routinely awards damages to parents when doctors negligently fail to diagnose or warn the parents of the possibility of genetically defective babies so that their conception can be prevented or fetuses aborted. While this may seem appropriate to many when they think of babies born with horrible diseases like Tay-Sachs or with deformities like anencephaly, what about parents who sue because their children were born with less severe afflictions like Down syndrome or deafness? At least one judge has refused to recognize suits for "wrongful birth" on the basis that they "can slide ever so quickly into applied eugenics":

> The very phrase "wrongful birth" suggests that the birth of the disabled child was wrong and should have been prevented. If one accepts the premise that the birth of one "defective" child should have been prevented, then it is but a short step to accepting the premise that the births of classes of "defective" children would be similarly prevented, not just for the benefit of the parents but also

for the benefit of society as a whole through the protection of the "public welfare." This is the operating principle of eugenics.[57]

Nor is eugenics completely rejected by scholars and academics. Dan Wickler, a prominent philosopher and bioethicist at the University of Wisconsin, seems to embrace the positive aspects of eugenics in a recent article entitled "Can We Learn from Eugenics?" in which he states:

> Today, a eugenic principle might call for prospective parents to screen pregnancies so that the children they bring to term have the greatest feasible genetic advantages. Tomorrow, these same parents might be encouraged to avail themselves of genetic interventions to cure and to enhance. Excepting perhaps the fetuses which are aborted as a result of such a programme, no one would be asked to make sacrifices. Because parents almost always seek advantages for their children—health above all—there is a congruence between a eugenicist's concern for the public and a parent's concern for his or her child. Where there is not, a voluntary programme would leave the decision to the parent.[58]

As Wickler observes, people want to improve their offspring genetically. But until now, the techniques of positive eugenics have been extremely limited. The cutting edge of this technology—our most sophisticated method of producing the greatest amount of genetic modification in the shortest time—has been embryo selection for implantation, that is, genetic testing combined with IVF, with only the two or three "best" embryos chosen to be implanted in the womb. But parents making these choices have been able to select only from naturally occurring genetic profiles, that is, from the natural DNA that the embryos inherit from reproductive cells or that result from occasional random genetic mutations. In short, it is just another of Lewis Thomas's "half-way technologies."

The process by which half-way technologies become whole technologies, like any journey of discovery, takes many steps. Consider the steps necessary to develop what are generally regarded as the most important whole medical technologies of the last century—penicillin and the polio vaccine—the years of research, the perfection of production processes and delivery vehicles. To be in a position to develop a whole technology for improving our genes, we first would have to understand the basic science of genetics. We would need to know the complex relationships between genes, proteins, the environment, and human characteristics. And we would have to learn how to manipulate DNA, to splice it in and out of people's genes, to turn it on and off.

But we are working on developing the means to do these things right now. Our knowledge of genetic science is growing exponentially. Gelsinger's death temporarily decelerated the process of gene therapy experimentation, but the demand for the whole medical technology of gene therapy remains irresistible. While investigational human gene therapy programs are focused on combating heritable disease, the genetic engineering techniques being perfected also will be able to alter non-disease characteristics. And the growing interest in behavioral genetics indicates that more and more scientists can be expected to turn their attention to exploring the relationship between genes and non-disease characteristics.

In short, we are poised at the very brink of transforming our ability to improve our genes from a half-way to a whole technology. The stage is now set for the fifth revolution.

4

The Fifth Revolution

When the human genome is completely unraveled, we will discover the genes not just for diseases or anti-social behaviors, but those that interact with the environment to produce all heritable physical and mental characteristics. Once the proteins associated with these characteristics are identified, we will be able to create drugs that mimic or suppress the actions of these genes, making it possible to produce short-term physical or mental changes. Genetic tests for non-disease traits will enable parents using assisted reproduction techniques like in vitro fertilization (IVF) to select embryos with the best combination of these genes. The same genetic engineering techniques used to prevent genetic disorders ultimately could be employed to install desirable non-disease characteristics in these embryos.

These techniques are the armamentarium of the fifth and final revolution in human genetics: the revolution in genetic enhancement. Some are still half-way, but some are whole, and all of them produce significantly more rapid and effective impacts than traditional patterns of human evolution.

The tools of genetic enhancement—drugs, genetic tests, genetic manipulation—will grow out of the tools of genetic medicine. In fact, enhancement techniques can be thought of merely as genetic medicine provided to "healthy" populations. The drugs that normal individuals may use to increase their brainpower, for example, will most likely be the same drugs physicians use to combat Alzheimer's and other cognitive impairments. This presents major challenges for those trying to regulate these enhancements, for manufacturers, health insurers, physicians, and for individuals deciding whether or not to use them.

In the first place, when is an intervention "enhancement" rather than "therapy"? This is important for a number of reasons. For example, health insurers may reimburse for genetic treatments to fight disease, but may consider enhancement to be like cosmetic procedures, which they do not cover.

Bioethicists and philosophers who have struggled with this question have proposed that an intervention be considered an enhancement when it produces an effect beyond that which is normal. But what is "normal"? For the most part, "normal" refers to some frequency of a characteristic in a population. We measure everyone's height, and call the average height "normal." But then is everyone who is shorter or taller than average "abnormal"? To avoid upsetting a lot of children and parents, child development specialists have adopted a statistical convention: people are abnormally tall or short if their height is more than two standard deviations from the population mean. This translates to about 5 percent of the population. Thus, a drug or genetic manipulation that resulted in a height greater than two standard deviations from the population mean would be considered an enhancement, while an intervention that moved a short person to within two standard deviations would be considered therapeutic. Similar conventions obtain for weight, IQ, and many personality traits. But these conventions are entirely arbitrary. Why should the line for abnormal height be drawn at two standard deviations from the mean, rather than one (in which case approximately 30 percent of the population would be abnormal) or three (virtually zero percent abnormal)? How is the relevant population defined? The normal height for a professional basketball player is much taller than the normal height for other persons. If the law prohibited employers from refusing to hire people because they were short, would a basketball team have to hire someone who was more than two standard deviations below the normal height for a basketball player to avoid being guilty of illegal discrimination? Conventions also can change as people alter their conceptions of what it means to be normal, and as the characteristics of the population change over time. If a lot of people used genetic enhancements to increase their height, the overall height of the population, as well as the mean, would rise. Families that were once considered to be of normal height might then find themselves abnormal.

In response to these conceptual problems, some philosophers have proposed that an intervention be considered an enhancement only when it produces a result, either in kind or degree, that goes beyond the bounds of the species.[1] A drug that allowed someone to grow a third arm would be an enhancement, as would a drug that increased a person's height to ten feet, but not a drug that only increased height, say, to seven feet. But this approach would call an intervention therapeutic even if used by a normal person, so long as the resulting change remained within the normal range for the species. This would produce the odd result that people five feet tall who became seven-footers would not be deemed to have enhanced themselves.

Apart from being arbitrary and unstable over time, population- or species-based conceptions of normality encounter other problems in attempting to distinguish clearly between enhancement and therapy. For example, the whole idea of immunization is to confer a degree of immunity from disease that goes beyond what people normally possess. Does this make vaccines enhancements? The NIH has approved gene transfer experiments to treat hypercholesterolemia, an inherited inability to clear cholesterol from the bloodstream. The treatment boosts the functioning of patients' low-density lipoprotein receptors beyond the normal range to compensate for their deficiency. If these experiments prove successful, could a health insurer refuse to pay for the treatments on the ground that they were not "therapeutic"?

Philosopher Eric Juengst has a better approach.[2] He suggests that vaccines and gene transfer treatments for hypercholesterolemia can be distinguished from true enhancements by their objective—to treat or prevent disease. He therefore offers the following guideline: An intervention is therapeutic when the object is to treat or prevent disease; otherwise it is an enhancement. Juengst's focus on "disease" rather than "normality" helps clarify the vaccine and hypercholesterolemia examples, but it isn't a complete solution to the conceptual problem. We still have to decide whether or not something is a "disease." Like the meaning of "normal," this too is a matter of convention to some degree, and it can fluctuate over time. Consider homosexuality, which until quite recently was considered a mental illness. But Juengst probably provides the best rule for defining the boundary between therapy and enhancement.

If Juengst's approach helps us to identify an enhancement, what makes an enhancement "genetic"? A change produced by genetic manipulation, such as by inserting genes into DNA, clearly qualifies. Arguably so does an enhancement achieved through assisted reproductive technologies coupled with genetic testing, for example, if a couple employing in vitro fertilization selected which embryos to implant based on the results of genetic testing for non-disease characteristics. We should even include performance- or mind-altering drugs whose discovery or manufacture are made possible by knowledge gained from the science of human genetics. Drugs that augment proteins whose role was uncovered as the result of genetic sequencing, and drugs manufactured with genetic engineering techniques such as recombinant DNA, should all be considered genetic enhancements.

Defined in this broad fashion, genetic enhancement is no longer just science fiction, but reality. Some athletes, for example, take a substance called erythropoeitin (EPO) to boost their strength and stamina. EPO is produced naturally in the body and increases the number of red blood cells, which carry oxygen. The more red blood cells there are, the more oxygen the blood can transport to the muscles and the better the muscles can function, which explains the athlete's interest. Not long ago, scientists began manufacturing synthetic EPO using recombinant DNA techniques. This makes it a genetic enhancement. Another enhancement drug is human growth hormone (HGH). This was originally harvested from the pituitary glands of cadavers, and was approved by the FDA for the treatment of a condition called pituitary dwarfism. The supply was extremely limited until drug companies began producing synthetic HGH using recombinant DNA. Once the shortage was alleviated and the hormone could be obtained by ordinary prescription, parents began asking physicians to prescribe it for children who did not suffer from pituitary dwarfism, but who were just short. There are even reports of parents seeking HGH for children who were exceptionally tall; the parents wanted to make them even taller to give them a better shot at a professional basketball career.

Drugs manufactured with recombinant DNA techniques are not the only forms of genetic enhancement currently available. A technique known as "pre-implantation genetic diagnosis," or PGD, is used in conjunction with in vitro fertilization to enable parents to avoid implanting embryos with serious genetic disease or abnormalities. As mentioned in the previous chapter, this technique is now being used to identify and avoid implanting embryos with late-onset genetic disorders like Huntington disease and certain forms of Alzheimer's. While parents employing IVF cannot yet decide which embryos to implant based on the results of genetic tests for traits like intelligence or beauty, they can and do make reproductive choices based on non-disease factors. For instance, egg and sperm donors are sometimes selected on the basis of non-disease characteristics. One couple ran an ad offering $50,000 to an egg donor who was tall, athletic, and had scored at least 1400 on her SATs.[3] As noted in Chapter 3, in 1976 a man named Robert Graham opened a sperm bank to collect and distribute sperm from geniuses.[4]

Another non-disease characteristic that enters into some reproductive decisions is gender.[5] Infertility clinics advertise a technique called sperm sorting in which the eggs are fertilized with heavier or lighter sperm depending on whether the parents want a female or male child. (As noted earlier, sperm cells with the male Y chromosome are lighter than sperm cells with the larger female X chromosome.) Some clinics are now using genetic testing to discern the sex of embryos before implantation. There are even reports of parents who abort fetuses because of their gender, and this practice is particularly widespread in countries like India where female children are regarded as a financial burden on parents.

But a handful of performance-enhancing drugs and gender-based reproductive decisions are a far cry from wholesale genetic tests for non-disease traits and from gene insertion and deletion for enhancement purposes, much less the creation of specialized human beings like the rescue workers in the scenario in the Introduction. This prompts some skeptics to insist that we have gone about as far as we can expect to go in the direction of genetic enhancement. Others, acknowledging the possibility of major breakthroughs, dismiss the likelihood that these breakthroughs will take place in the near future. For the time being, they say, the fifth revolution in human genetics deserves to remain firmly in the realm of science fiction, rather than sharing the stage with other genetics issues in debates over ethics and science policy.

This would be dangerously shortsighted. The pace of genetic progress—particularly the development of faster sequencing machines and more sophisticated computer programs to understand the interactions of multiple genes as part of the revolution in genetic information, and the perfection of gene transfer technology growing out of the therapeutic revolution—signifies that these skeptics are wrong. Remember, the techniques of genetic enhancement are the exact same techniques as the ones used in genetic medicine, and the demand for progress in genetic medicine is huge. An enormous commitment of resources and brainpower is being made to speed the discovery of tests for genetic disorders and cures for genetic diseases. Moreover, as emphasized earlier, these efforts are the result of the unprecedented confluence of two powerful forces of discovery: genetic science and cybernetics. The import is clear: Even if no one were interested in doing research on genetic enhancements, the advances in genetic medicine would pave the way.

But plenty of people, droves of venture capitalists and entrepreneurial geneticists, for example, will be interested in enhancement research. The demand for genetic enhancements ultimately will be at least as large as the demand for therapeutics. In fact, the commercial value of whole rather than half-way technologies for improving human capabilities may be virtually unlimited. Just as it is said that there is no such thing as being too rich or too thin, there may be no such thing as being too handsome or too intelligent. True, there has been little progress so far in identifying genes for non-disease traits, and no one yet appears to be attempting to manipulate genes to achieve enhancement objectives. The first focus of research understandably will be on genetic illness, especially since that is where the federal government is injecting its huge amounts of research funding. But it is clearly only a matter of time before enough of the therapeutic puzzles in human genetics are solved and enterprising researchers turn their attention to non-disease genes.

Another group of skeptic concedes that genetic enhancement technologies are bound to develop eventually, but argues that they will have little impact on society. Enhancement is nothing new, they assert. People always have tried to improve themselves or their children in various ways. Education is an obvious example; many parents opt for private over public schools despite the considerable additional cost. Similarly, people go on diets, work out at fitness centers, sign up for speed-reading and memory-improving courses. They try to marry "upward," seeking mates who will increase their employment opportunities, social standing, and wealth. They push their children into extracurricular activities and punish them when they don't spend enough time practicing the piano or the violin. They even go to extremes. One woman who wanted to free up a spot on the junior high school cheerleading team for her daughter was sentenced to ten years in jail after she offered to give her diamond rings to a hit man if he murdered the mother of one of the other cheerleaders.[6] In order to increase their toddler's chances of being accepted at an exclusive preschool in Manhattan, one couple repeatedly fed the child okra prior to her admissions interview; the parents anticipated that she would be asked to name her favorite foods, and hoped that answering "okra" would give her an advantage.[7]

Many current efforts at self-improvement employ biological interventions. People take foods and drugs to improve their athletic and cognitive performance, from carbohydrate loading to power bars to caffeine. They subject themselves to painful and expensive cosmetic surgery to improve their appearance, including breast augmentation and reduction, liposuction, rhinoplasty (nose jobs), and face-lifts. They pay thousands of dollars for their children's orthodonture. Between 1992 and 1997, the number of persons obtaining liposuction rose 215 percent.[8] Some of these self-improvement efforts, like "marrying upward," can even be said to be a type of germ line enhancement, since they have an effect on the genetic makeup of future generations.

Given this evidence of our age-old determination to improve ourselves, the skeptics might ask, what is so special about genetic enhancement? Society has managed to cope with the effects of these other forms of self-improvement. The Olympics screen athletes for blood doping, and some beauty pageants make contestants swear they have not had cosmetic surgery. Why should we be particularly concerned when self-improvement involves genetics?

The answer lies in the enormous potential power of genetic technology to transform society. In the first place, genetic enhancement could be substantially more effective than current means of self-improvement. At present, people can alter their weight by diet and liposuction, and they can use cosmetics and cosmetic surgery to alter their appearance. But the improvements made by taking speed-reading, memory, and social skills courses, or going to gyms to build muscles, are fairly modest, and it usually takes constant practice to maintain them. Moreover, these interventions rarely stray from "normal" population ranges for physical characteristics, and with the exception of cosmetic surgery, the changes are usually transitory. Even athletes who take performance-enhancing drugs like steroids gain only a limited effect; weightlifters who take steroids, for example, can bench press an additional eight kilograms, or about seventeen pounds.[9] Blood-doping, intravenously infusing blood to increase oxygen-carrying capacity, is more dramatic, producing a 15 to 30 percent increase in endurance.[10] Amphetamines improve performance by 3 to 4 percent in shot putters, 1.5 percent in runners, and 0.6 to 1.2 percent in swimmers.[11] These enhancements may be enough to win competitions, but they are not transformational.

Genetic interventions, on the other hand, could produce far more profound and long-lasting effects. Instead of dieting to lose or gain weight, people might be able to alter their metabolisms in fundamental ways. Increases in strength, stamina, and endurance could go far beyond population norms, and even beyond world-class athletic norms. Cognitive enhancements could produce virtually unlimited increases in intelligence, memory, and other cognitive abilities. Traits important for social success, such as charisma, a sense of humor, cheerfulness, or creativity, might prove amenable to significant genetic manipulation.

But genetic enhancement conceivably has a more ambitious goal than even these remarkable achievements. Its Holy Grail—the ultimate enhancement—would be to conquer the aging process itself. In the short term, genetic science is likely to give rise to drugs and other medical interventions that combat the diseases of old age, both chronic and acute. But genetic engineering may make it possible to forestall the aging process at the cellular level. If this process could be interrupted, cell life might be prolonged indefinitely. Interrupt the process at a young enough age and you might even produce eternal youth.

Even if this ambitious goal is not attained in the foreseeable future, genetic enhancements could still differ significantly from traditional forms of self-improvement. Genetic enhancement could affect a greater number of characteristics at one time. People mostly work on one or two traits—their facial appearance, their weight, their ability to memorize facts, and so on. Taking on too many challenges, they are told, is a recipe for failure. Cosmetic polysurgery, altering many traits at once, or involving different parts of the body, is extremely rare.[12] Even so-called "makeovers," visually startling as they may appear, are achieved largely just by changing makeup, hairstyle, and clothing.

Genetic enhancement, on the other hand, could be wholesale. People could take an assortment of somatic enhancements—pills or infusions of genes, for example—to change a host of characteristics simultaneously. The only inherent limits on somatic enhancements would be their cost, technical problems of production that restricted supply, and interactions between enhancements that reduced effectiveness or produced unwanted side effects. Embryo selection for enhancement purposes—implanting into the womb only those embryos that genetic testing had shown to have the most desirable combination of genes—would be limited only by the need to develop tests for non-disease characteristics and by the naturally occurring frequency of the preferred genetic combinations. With gene transfer technology, the number of traits that might be altered could become virtually unlimited.

By permitting individuals to alter dramatically a large number of traits, genetic enhancement could give them enormous, perhaps decisive, advantages over others. Current forms of self-improvement can change appearance, although rarely dramatically (consider how unusual a case like Michael Jackson's is), together with some aspects of performance and some degree of cognitive functioning. But genetic enhancement could make a single individual stronger, taller, more beautiful, more intelligent, more charming, more creative, more fun to be with. Consider the following description by political philosopher Michael Walzer:

> Here is a person whom we have freely chosen . . . as our political representative. He is also a bold and inventive entrepreneur. When he was younger, he studied science, scored amazingly high grades in every exam, and made important discoveries. In war, he is surpassingly brave and wins the highest honors. Himself compassionate and compelling, he is loved by all.[13]

Walzer points out that wholesale changes like these would create advantages not just in one or two realms of social activity, but in a whole range of social endeavors. We occasionally see polyperformers, but they are rare and their advantages typically are limited to one or two spheres, like politics and wealth or celebrity. Genetically enhanced individuals could cross "spheres of distributive justice," to use Walzer's term. They could dominate not just one or two forms of social competition, like running for political office, being a highly paid actor, or having great wealth, but all of them.

The ability of enhanced individuals to obtain advantages in all spheres of justice might not be a great problem if genetic enhancement were available to everyone. Competition for societal benefits simply would take place on a higher plane, in much the same way that tennis changed when players began using larger rackets, or pole-vaulting heights increased with the introduction of fiberglass poles. Natural talent would still differentiate people. If genetic enhancement enabled everyone to gain an extra couple of feet in height, for example, you would still have to be naturally tall to become a professional basketball player, but "naturally tall" might mean eight or nine feet. Genetic enhancement might add muscles, but hard work would still pay off.

Yet genetic enhancement is not likely to be available to everyone. The price of somatic enhancements—genetically engineered drugs, for example—will be high, at least when they are first introduced, so that their manufacturers can recoup their research-and-development costs and reap the financial benefits of the exclusive marketing rights conferred by their patents. Prices may go down over time, but may still remain relatively steep. Growth hormone therapy for a twenty-kilogram child still cost $14,000 per year in 1998, more than a dozen years after researchers figured out how to synthesize the product using recombinant DNA.[14] The most effective and long-lasting forms of genetic enhancement will involve IVF, which costs approximately $37,000 per delivery—without the added cost of genetic enhancement. Currently only about 40,000 couples in the United States take advantage of this form of assisted reproduction.[15] While this number would increase because of the demand created by genetic enhancement, many people will still find the cost prohibitive.

In contrast, current types of self-improvement are much more widely available. More than twenty million Americans belong to commercial health and sports clubs.[16] Almost half a million per year receive liposuction, breast augmentation, eyelid surgery, skin resurfacing, or face-lifts.[17] Costs, while beyond the means of the poor, are generally below $5,000.

A more profound difference between traditional forms of self-improvement and genetic enhancement is that enhancements may be able to alter the germ line. Somatic enhancement in adults can go only so far: Most physical characteristics and non-disease traits are largely fixed by adulthood. Somatic enhancements might be able to alter adult weight and performance, but they are unlikely to be able to radically change characteristics like facial appearance or height. Moreover, somatic enhancements most likely will be temporary, requiring repeated doses to maintain their effect. More dramatic changes in adults might be accomplished by inserting altered genes, analogous to gene therapy. But the adults' natural DNA would still control many functions.

Since the bodies and minds of children develop more radically than those of adults, genetic enhancements might be more effective if introduced at a younger age. Human growth hormone, which has no apparent effect on the height of adults, does increase the height of young children (although as it turns out, children without pituitary deficiencies don't end up taller than they would be without the hormone, they just reach their natural height more quickly).

Even more significant enhancement effects might be achieved in fetuses, especially at early stages of development, with the most significant changes being produced before the fetal cells have become highly differentiated, so that the enhancement reaches all of the relevant areas of the body. Introduce altered genes at an early enough stage of fetal development, and you have the enhancement equivalent of the gene therapy experiment proposed by French Anderson and Emil Zanjani that was described in the preceding chapter, in which they acknowledged that the altered DNA could find its way into the cells that would become eggs or sperm, and therefore be passed on to succeeding generations. In short, you have germ line genetic enhancement.

Even more dramatic germ line enhancement could take place if enhancement accompanied in vitro fertilization. An egg could be fertilized in the laboratory, and when it had divided into about eight cells, one of the cells could be removed and its DNA manipulated to produce enhancement effects. The cell would then be induced to begin dividing, creating a new, enhanced embryo, which could be implanted in the womb and brought to term. Through the mechanism of cell division, the enhanced DNA would be deposited in all of the nuclei of the child's cells, including its reproductive cells, so that when the child grew up and had its own children, the enhanced DNA would be transmitted to them via the child's sperm or eggs.

In contrast to traditional forms of self-improvement, which produce only modest and indirect effects on future generations, germ line genetic enhancement would pass the social advantages of enhancement from one generation to the next. The children of today's celebrities and families of wealth certainly enjoy advantages that other children don't. The reason that so many famous actors are the children of famous actors, for example, is likely to be due as much to their parents' connections as to talent. But the much more dramatic and far-ranging advantages produced by genetic enhancement would dwarf the privileges currently afforded by one's lineage, except perhaps in those societies that still accept the concept of inherited nobility.

The skeptics who hold that the prospect of genetic enhancement presents no different or greater challenges for society than existing self-help methods therefore are clearly wrong. Genetic enhancement is likely to affect a much broader range of traits and produce far bigger improvements. It is liable to alter a much larger number of characteristics at once. It will be more expensive, limiting access to wealthier individuals and families. Those who became enhanced most likely would enjoy enormous advantages over others, advantages that confer privileges upon them across the entire spectrum of social spheres. Germ line genetic enhancement would enable these advantages to be passed on to their children. Genetic enhancement might even significantly lengthen one's lifespan, perhaps even give people eternal youth.

Nothing we can now do to improve ourselves even comes close.

But we haven't considered the most momentous potential consequence of genetic enhancement, the greatest challenge it presents. The germ line changes introduced as a result of genetic enhancement might produce individuals who deviated so substantially from current population norms that the result would be a new species of life on the planet.

A species that is far stronger, handsomer, and more intelligent, and that quite possibly would live forever.

A species that is no longer human.

The skeptics rise once more to their feet. Even if some people became genetically enhanced, they argue, this won't alter the species. In a 1999 article in the prestigious journal *Science,* Professor Jon W. Gordon stated that the notion that genetic enhancement would enable us to control human evolution is "totally without scientific foundation," and that the impact of gene transfer on the species "would be negligible."[18] The reason, Gordon asserts, is the minute effect that genetically modifying a few individuals would have on the frequency of genes in the entire population:

Every month worldwide approximately 11 million babies are born. The addition of one genetically modified individual could not significantly affect gene frequencies. Moreover, if the "enhanced" individual had his or her first child at the age of 20, then 2,640,000,000 unengineered children would be born during the interval between the birth and procreation of the gene recipient. Even if 1000 successful gene transfers were performed per year, a number not likely to be achieved in the foreseeable future, those newborns would constitute only 1/132,000 of all live births. Thus, any effort to enhance the human species experimentally would be swamped by the random attempts of Mother Nature.

Gordon is correct that genetic enhancement that was not available to everyone would not dramatically change the genetic composition of the bulk of humanity. But we are not assuming that genetic enhancement will change everyone. Gordon's calculations merely underscore the degree to which an enhanced group might appear to be, and think of themselves as, divergent. The human species will persist, but it will no longer be alone.

The potential of genetic enhancement, in short, is immense, and the challenges it creates are unprecedented. To determine if they can be met, we now need to examine them more closely.

5

Safety and Effectiveness

As we saw in the previous chapter, genetic enhancement potentially could produce major changes in human abilities and characteristics. Yet some skeptics, we know, disagree. For them, all this
talk about genetic enhancement is just hype. Enhancement simply won't work, they say. It'll be ineffective. A dud.

The question is: How will we know if they are right? In fact,
how will anyone know if genetic enhancement works? Well, how
do we know that any biological intervention, like taking a drug or
having surgery, is effective?

For many medical interventions, the answer is easy. We identify
a set of objectives (call them endpoints), administer a drug or procedure, and ascertain if the endpoints are met. Depending on the
circumstances, these endpoints might be reducing a fever, preventing immediate death or prolonging survival, clearing a blocked
artery, or lowering the amount of HIV in a person's bloodstream.
These endpoints can be measured by fairly objective methods: ther-
mometer readings, blood flow, lab tests.

We might also be able to identify clearly defined endpoints for a number of genetic enhancements. Strength can be measured by how much weight people can lift, endurance by how far they can walk or run, visual acuity by the size of the letters they can read on an eye chart. Tests even exist for memory and intelligence, although IQ, the principal measure for assessing intelligence, is controversial, and in the case of memory, the endpoint, and therefore the appropriate tests, might differ depending on the goal, for example, increasing the amount of information someone can remember versus the duration of recall. Height at first seems easy: all you need is a tape measure. But as the experience with human growth hormone (HGH) described in the previous chapter demonstrates, a height enhancement might increase a person's ultimate height or merely hasten when they attain their natural height.[1] Would only one of these be considered evidence of effectiveness, or both?

Other potential enhancement endpoints are less well-defined and more difficult to measure. How would we assess beauty or charisma, for example? In the case of beauty, we might employ the same technique as cosmetic surgeons: allowing patients to select a representation of their desired nose or breast shape and comparing it with the surgical outcome. For charisma, we would need some scale that captured the effect that the enhanced person had on others.

Even if there were some way to measure the relative physical or mental changes produced by genetic enhancement, however, this might not reveal enough to enable a person to determine effectiveness. The goal of the enhancement arguably is not just to increase IQ or strength, but to enable the individual to get into Harvard or to win an Olympic weightlifting medal. Therefore, measuring short-term or immediate effects may not tell us whether or not the intervention achieves its long-term objectives. Instead, we will have to identify the appropriate long-term goal and evaluate the impact of the enhancement in terms of its achievement. At a minimum, this means there is likely to be a significant delay between being enhanced and obtaining the benefit, and therefore in discovering whether or not the enhancement actually works. Moreover, ultimate goals might vary from person to person. This would make it difficult to assess effectiveness because it might be difficult to combine results from different people to obtain an overall result. Without a large number of observations of the effects of an enhancement in many people, it is difficult to attain the statistical confidence to rule out that a result occurred purely by chance, rather than because of the intervention. Some people may still get into Harvard without being genetically enhanced.

While these difficulties will plague future efforts to assess the effectiveness of genetic enhancements, they are not new. We have encountered them before in currently existing biomedical interventions. Take cosmetic surgery. When they remove the bandages, patients may be able to see if they ended up with the nose or breasts they wanted, and their self-image may improve abruptly. But they will still have to wait to find out if they become more popular or if their sex life improves, and the identification of these endpoints and the determination of whether or not they have been achieved will be highly subjective.

Nor are these difficulties in assessing effectiveness limited to cosmetic medicine. There are no agreed-upon objective measures for evaluating pain, so the effectiveness of pain-reducing medications must be measured by the subjective responses of patients. Bioethicist Baruch Brody describes another example, the case of "clot-busters"—drugs used to treat heart attacks by breaking up blood clots in the coronary arteries.[2] One way to measure whether or not they work is to determine if in fact they break up blood clots. But breaking up the clots is not the real endpoint; the real objectives are to reduce the severity of the heart attack, prevent a recurrence, ease discomfort, reduce recovery time, restore functionality, and ultimately prolong life. Drug manufacturers, eager to gain rapid approval of their new products, wanted the Food and Drug Administration to use clot-busting per se as the endpoint for deciding whether or not to approve their license applications. The industry argued that it was the most objective and easily measured effect. But the agency held out for more long-term and subjective effectiveness data. Even then, there is no objective way to assess the key endpoints of restoring functionality or prolonging life. The value of restored functionality may vary from one person to another depending on how active they wish to be. The value of an extra year of life may vary as well, particularly if for some people it may be a year filled with immobility or pain.

While evaluating the effectiveness of biomedical interventions in general, and of genetic enhancements in particular, is difficult, it is critical. For one thing, without knowing how well something works, how will people decide how much to pay for it? How will they be able to winnow the truth from advertising claims? And how will they choose between alternatives—a somatic enhancement drug, an infusion of altered genes, a germ line transfer?

But having good information about how well a genetic enhancement achieves its endpoint is essential for another, arguably far more important, purpose: to enable people to compare the benefits of an intervention with its risks. From the perspective of persons considering whether or not to become enhanced, the key factor will be the *net* benefit expected from the enhancement: Does it produce more good than harm, and if so, how much. To know this, they will need to know both how effective it is, and how safe. Will an enhancement drug have side effects? If so, how serious will they be? Will an infusion of altered genes interfere with other biological functioning? Will inserting or deleting DNA throw off the complex interaction of other genes? The very power that makes genetic technologies so attractive carries with it the threat of grave harm.

Scientists have long been troubled by the potential dangers of genetic manipulation. The first big technological breakthrough, recombining DNA, raised concerns at an early research conference in 1973. It led to a call from a committee of the National Academy of Sciences, chaired by leading geneticist Paul Berg and including James Watson, to convene an international meeting to discuss safety issues raised by "the creation of novel types of infectious DNA elements whose biological properties cannot be completely predicted in advance":

> There is serious concern that some of these artificial recombinant DNA molecules could prove biologically hazardous. One potential hazard in current experiments derives from the need to use a bacterium like *E. coli* to clone the recombinant DNA molecules and to amplify their number. Strains of *E. coli* commonly reside in the human intestinal tract, and they are capable of exchanging genetic information with other types of bacteria, some of which are pathogenic to man. Thus, new DNA elements introduced into

E. coli might possibly become widely disseminated among human, bacterial, plant, or animal populations with unpredictable effects.[3]

The meeting, attended by 150 scientists, four lawyers, and sixteen journalists and held at the Asilomar Conference Center in California in 1975, focused on developing safety guidelines for conducting recombinant DNA research.[4] Concerns persisted despite the guidelines, which were soon adopted by the NIH. Experts were particularly worried about the impact on plants and animals if researchers inadvertently released recombinant DNA into the environment. But as the researchers gained more experience with the new technology, their fears lessened. In 1978, the guidelines were relaxed.[5]

Safety issues again loomed large when researchers in 1997 announced that they had successfully cloned a mammal from an adult cell. The birth of Dolly the sheep was only tangentially related to the development of genetic enhancement techniques; its significance lay mainly in raising the possibility that humans could be cloned for reproductive purposes and for purposes of producing transplant organs that would not be rejected by their recipients. But Dolly was a dramatic breakthrough nevertheless, and it stunned the skeptics who had maintained that genetic enhancement would never become a reality.[6] Now that Dolly had undermined their conviction that genetic enhancement was impossible, they seized upon the risks associated with her birth. Dolly's gross anatomy— her physical appearance—seemed normal. But she was the only survivor of 277 activated eggs; if these results carried over to humans, it would take 277 human eggs to create one cloned human embryo. Where would all these eggs come from, the critics asked. Moreover, Dolly's telomeres—the structures at the ends of the genes that grow shorter as an organism ages—were not the length of a newborn's, but the length of her mother's.[7] This suggested that Dolly's genes were "aged" at her birth and that she would not live long. Questions also were raised about whether Dolly could reproduce. More recently, she appeared to have developed arthritis prematurely.[8] But other cloned mammals, such as cows, have grown and reproduced normally.[9]

At the same time, gene therapy experiments in humans continued. The successes with Ashanti DeSilva described in Chapter 3 led to efforts to introduce corrected genes into patients suffering from a range of diseases, including other auto-immune disorders, cystic fibrosis, and inherited forms of hypercholesterolemia (high cholesterol). All of the experiment seemed to be proceeding safely until the death of Jesse Gelsinger in 1999, described in Chapter 3. The risks inherent in attempting genetic enhancement interventions in humans once again loomed large. While the experiment that killed Gelsinger was aimed at developing a gene therapy, the same techniques would be employed to achieve genetic enhancement.

Particularly troubling is the risk of serious side effects from attempts to genetically enhance children. Conscientious scientists would first test the interventions out on animals, but successful laboratory and animal experiments only prove so much. Many medical interventions that seemed safe in these so-called preclinical investigations proved dangerous in human use. For example, the FDA approved the drug Rezulin in 1997 with exceptional speed, despite animal studies and clinical trials revealing rare signs of jaundice, and withdrew it from the market in 2000.[10] What if instead of turning them into brilliant, wonderful adults, attempts to genetically enhance children made them sick, or even, as in Jesse Gelsinger's case, killed them?

Similar fears beset the idea of enhancing early-stage fetuses or embryos. As noted in Chapter 3, these techniques might offer the best chances of significantly improving offspring characteristics, since all cells would incorporate the enhanced DNA. But what if the enhancements proved lethal, so that the embryos or fetuses died, like the 276 eggs that failed in the experiment that eventually yielded Dolly? What if enhancements backfired, and the embryos or fetuses became monstrosities before they died? Or worse, what if the monstrosities were born alive?

In the face of these uncertain dangers, the question is how much freedom people should be accorded to try to enhance themselves or their offspring. One approach might be to let people decide for themselves whether or not to accept these risks. Genetic enhancement would be treated like most other forms of self-improvement. Just as people choose to accept the risks of injury when they engage in athletic or fitness programs, they could decide whether or not to accept the risks of genetic enhancement. In making these decisions, people would need access to valid and reliable information about risks and benefits, which they could obtain from the companies and professionals who provide enhancement services. But the ultimate decisions would be left up to the individual.

Yet, as developed in Chapter 4, genetic enhancement is not like traditional methods of self-improvement. For one thing, it is potentially far more powerful. And this power makes it potentially far more dangerous. Genetic enhancement may produce much greater effects, but at a much greater risk of serious injury.

The appropriate model for making decisions about whether or not to enhance oneself may not be the model for deciding whether or not to work out or go jogging, then, but for deciding whether or not to take a powerful prescription drug or to be probed by a new invasive medical device. These decisions are not left solely to the individual. They are made in the first instance by government regulators, who decide whether or not a drug or device may be lawfully marketed and what claims it can make. Only then is the decision about whether or not to accept the risks in view of the expected benefits turned over to the individual, in consultation with health providers. The question, then, is what is the proper role of government regulation in controlling access to genetic enhancements.

From the outset, scientists working to develop genetic technologies recognized the need for some government oversight. The letter from Paul Berg that led to the Asilomar Conference in 1975 called for a worldwide moratorium on recombinant DNA experiments until suitable research guidelines could be implemented. The moratorium appears to have been effective, and was only relaxed some months later when the National Institutes of Health created a Recombinant DNA Advisory Committee, known as the "RAC," to come up with safety rules for DNA research. The first set of guidelines, adopted in 1976, required all large-scale, NIH-funded recombinant DNA experiments to be approved by the RAC.[11]

The RAC originally had twelve members, all of them scientists.[12] In 1978, it was expanded to twenty-five members, with one-third of those to include lawyers and bioethicists, designated "public members." In 1982, a report entitled *Splicing Life* was issued by the President's Commission for the Study of Ethical Problems in Medicine and Biomedical and Behavioral Research. The President's Commission, best known for the 1979 *Belmont Report,* which laid out the ethical rules for human subjects research, called for expanded consideration of the ethical and social implications of gene therapy. However, the report was broadly supportive of the technology.

In response to the report and to congressional hearings, the RAC and the NIH began to consider whether the RAC, which had to that point focused on laboratory and animal experiment using recombinant DNA techniques, should review proposals for human gene therapy experiments. In 1984, it established a Human Gene Therapy Working Group, chaired by a bioethicist, LeRoy Walters, and comprising three laboratory scientists, three clinicians, three lawyers, three ethicists, two public policy specialists, and a representative of the public. This group proceeded to issue a 4,000-word document entitled "Points to Consider" to provide guidance to researchers seeking RAC approval for federally funded human gene therapy experiments.

The "Points to Consider" noted that the NIH was prepared to approve funding of human experiments if they presented "no significant risk to health or the environment."[13] The approach employed by the NIH to assure the absence of significant risk followed the approach generally used in the United States for biomedical research on human subjects. In addition to the RAC, which determined whether or not the experiment was safe enough to merit government funding, each institution conducting the research, such as a hospital or university, had to have its own internal Institutional Review Board (IRB). The IRB was charged with reviewing and approving the study plan or "protocol" prior to the start of the experiment, making sure that subjects received adequate information about potential risks before agreeing to be part of the study, and monitoring the progress of the experiment to identify and respond to any safety problems that arose. (Recall that, as mentioned in Chapter 3, Martin Cline failed to wait for this IRB approval when he attempted gene therapy in the patients in Israel and Italy in 1980, and was severely punished as a result.)

In 1985, the RAC issued a revised "Points to Consider" in which it signaled its willingness for the first time to consider approving protocols for human gene therapy experiments. But the federal government's Food and Drug Administration was also interested in regulating these experiments. The FDA regarded gene therapy as falling within the categories of drugs, medical devices, or biologics, which the FDA regulates under the Federal Food, Drug, and Cosmetic Act. Moreover, the agency was keen to play a key role in the exciting new field of genetic medicine. In a policy statement issued at the end of 1984, the FDA staked out its territory. It also announced what its position would be on how these new technologies should be regulated: "The Agency possesses extensive experience with the administrative and regulatory regimens described as applied to the products of biotechnological processes, new and old. . . . Nucleic acids [DNA] used for human gene therapy trials will be subject to the same requirements as other biological drugs."[14]

This was a remarkable approach for the FDA to take. Despite widespread perceptions that genetically modified drugs and biologics, not to mention altered human DNA itself, presented special safety concerns, the agency was making it clear that it would not subject these technologies to any greater scrutiny than any other product under its jurisdiction, nor require them to undergo any special safety testing. The agency's position undoubtedly reflected the by then considerable safety record of recombinant DNA techniques. It was surely a boon to the fledgling biotechnology industry. But it created unease in many circles, particularly in Europe, where opposition arose especially against the FDA's attitude toward genetically modified food, which it proposed to regulate and label the same as "naturally grown" products.

The FDA's regulatory requirements basically obligate sponsors to obtain FDA approval before they conduct human experiments, and to secure a license from the agency before they market any commercial products stemming from the experiments. In order to secure permission to conduct a human experiment, the sponsor of the study must submit data from laboratory and animal studies that convinces the FDA that it would be safe to test the intervention in humans. To obtain a marketing license, the sponsor must submit data from well-controlled human studies that demonstrate that the product is safe and efficacious. As a condition of the license, the manufacturer also agrees to accept the agency's decisions about what information may appear on product labeling and in advertising.

As the NIH embarked on the Human Genome Project, the FDA continued to exert its regulatory authority over gene therapy in tandem with the RAC. In 1991, the FDA issued its own "Points to Consider" document, and in 1992, it created a new Office of Therapeutics Research in the FDA's Center for Biologics Evaluation and Research (CBER). A new oversight unit for human gene therapy, the Division of Cellular and Gene Therapies, under the direction of Philip Noguchi, was also created within the CBER. By the mid-1990s, gene therapy researchers and the biotechnology industry were becoming increasingly frustrated with what they viewed as duplicate regulatory programs by the FDA and the RAC. In order to proceed with a human gene therapy experiment that was funded by the NIH, sponsors had to seek approval from two bureaucracies, using two different application processes. Despite efforts to coordinate the approval processes, industry discontent continued to mount. Finally, in 1996, the director of the NIH, Harold Varmus, announced his intention to cede regulatory authority to the FDA.[15] Beginning in 1997, the FDA gained exclusive regulatory authority over human gene therapy experiments and products. The RAC was reduced in size and relegated to a largely advisory and educational role.

With the FDA now in charge of regulating gene therapy and gene therapy experiments, the agency is poised to regulate genetic enhancement as well. Somatic enhancements like drugs made with recombinant DNA would fit within the definition of a "drug" in the Federal Food, Drug, and Cosmetic Act, which includes "articles . . . intended to affect the structure or function of the body of man,"[16] whether or not they are aimed at diagnosing, curing, or treating disease. Enhancements achieved by infusing enhanced genes into the body or by gene insertion or deletion would resemble gene therapy, over which the FDA also has asserted jurisdiction. Even the medical equipment used to produce and deliver enhancements, such as chemical reagents, genetic sequencing machines, and implements like micro-pipettes and syringes, would fall within the definition of "medical devices" under the act. The only product category under the agency's authority that would not include genetic enhancement is "biologics," viruses and simi-

lar products which, by the terms of the act, are for "the prevention, treatment or cure of diseases or injuries of man."[17]

In regulating genetic enhancement, the FDA undoubtedly would
call upon its track record for cosmetic devices such as liposuction
machines (when removing fat for cosmetic purposes rather than
to avert the health effects of morbid obesity), non-prescription
contact lenses, and breast implants used for non-reconstructive
purposes. But the FDA's experience with these cosmetic devices
recalls many of the difficulties outlined earlier in this chapter.

In the case of liposuction, the government panel responsible for
assessing effectiveness ultimately had to admit that the only true
indicator of how well the machines functioned was not how much
fat they removed but the subjective measure of how satisfied patients were with the results.[18] When it came to breast implants, the
agency's concern in 1992 over the safety of gel-filled silicone implants prompted it to limit their use to research studying their
safety and efficacy only for reconstructive purposes—that is, in
women who had had breast cancer surgery, severe injury to the
breast, a birth defect affecting the appearance of the breast, or a
medical condition causing a severe breast abnormality.[19] The implants could not be used, much less studied, for cosmetic breast
augmentation. At the same time, though, the FDA made no distinction between cosmetic and reconstructive uses for saline breast
implants: both were permitted. This signifies either that the agency
felt that the risks posed by saline implants, as opposed to silicone-
filled implants, were so small that they were outweighed by cosmetic as well as therapeutic benefits, or, more likely, that the agency
simply did not come to grips with the significance of the enhance-
ment/therapy distinction for these products. In the case of contact
lenses, the FDA again treats corrective and cosmetic lenses the
same for regulatory purposes. Manufacturers who wish to market
contact lenses that do not improve vision, but merely change the
color of the iris or impart odd visual effects like cat's eyes, must
meet the same safety and efficacy requirements as manufacturers
of corrective lenses. At first this may seem appropriate, since the
risks to the eye of the wearer, such as the risks of infection, presumably are similar. But remember that safety is a relative concept. Nothing is absolutely safe; the question is whether the potential risks are outweighed by the potential benefits. In the case

of cosmetic contact lenses, even though their risks are the same as those from corrective lenses, their benefits are far less significant, since they do nothing to correct vision. Given that the risks from any type of contact lens are not trivial, the FDA should require greater proof of safety for cosmetic lenses than for corrective lenses. The agency's failure to do so suggests that it might make similar mistakes when it attempts to regulate genetic enhancement.

Another weakness in the FDA's ability to regulate genetic enhancement is its lack of jurisdiction over the practice of medicine. The FDA has asserted its authority over gene insertion and manipulation techniques such as human cloning, but its claim is strongly disputed and has not yet been upheld by the courts.[20] Technically, the agency has authority over drugs and medical devices, but not over the way health care professionals prescribe or wield them.[21] In theory, then, the agency lacks the ability to regulate the use of genetic testing for enhancement rather than therapeutic purposes.

More importantly, the FDA does not regulate the practice of IVF. This is a particularly significant loophole since, as mentioned in Chapter 4, IVF is a gateway technology for many forms of genetic enhancement. Without IVF, you cannot test and select embryos for enhancement or employ germ line and other types of gene transfer enhancement techniques. The lack of FDA oversight over IVF would not be so glaring if some other government agency had regulatory authority in its stead, but that is not the case. Except for a couple of state programs and a federal law that requires clinics to report their success rates, IVF basically is not regulated by the government at all.[22]

Another regulatory loophole applies to egg and sperm donation. Neither the FDA nor any other federal agency oversees the practice, even when it is conducted by a fertility clinic or commercial enterprise. State regulation is also extremely limited.

This means that, at the least, a whole range of genetic enhancements—the passive forms of enhancement described in Chapter 4 such as embryo selection in the course of IVF and abortion decisions for enhancement purposes—would be beyond the FDA's purview. IVF physicians and clinics, for example, would not have to conduct studies to prove to the government that their genetic enhancement interventions were safe and effective. As a consequence, safety and efficacy information about these technologies would be hard to come by, despite its importance for people contemplating whether or not to enhance themselves or their children.

Yet some data already exist, and they indicate that there indeed may be safety hazards. Studies in Australia recently reported that babies born after in vitro fertilization (IVF) had a higher risk of birth defects and of being underweight.[23] One in ten infants born through IVF had birth defects, twice the normal rate. The low birth weight problem may be because IVF usually results in several embryos being implanted in the womb and sometimes all survive to term; multiple births are known to increase the risk that the babies will be underweight.

By far the greatest impediment to FDA oversight over genetic enhancements, however, is that they are likely to become available as unapproved uses of FDA-approved products and services. Consider a drug that enhances cognitive functioning, such as memory or intelligence. How will it be discovered? Perhaps by accident, but if its discovery takes place in the course of an organized scientific effort, most likely it will be by scientists searching for a treatment for Alzheimer's disease or one of the other cognitive impairments that accompany the aging process. Imagine for the moment that you are the head of the company that made the discovery, and now you want to market it. Since the product is a drug, you need FDA approval. The agency is well known for prolonging its approval process, ostensibly to make sure that drugs are safe, and it only acts favorably with dispatch when it is asked to approve a breakthrough drug to treat a serious medical condition. So as the manufacturer, are you going to test and seek FDA approval for an enhancement claim, or for a claim to treat a devastating, widespread illness like Alzheimer's disease? Why would you not select the smoother pathway to FDA approval, which even under ideal circumstances can take years and costs millions of dollars?

But, you object, you don't want to sell the drug only to Alzheimer's patients. Your company wants to meet the potentially even larger demand for enhancement use—to sell the drug to people who want to function better at work, for example, or to students who want to ace their exams. If the FDA approves the drug only for use in Alzheimer's patients, you point out, the company won't be able to tap this vastly bigger market.

But you would be incorrect. There is nothing in the laws administered by the FDA that makes it illegal for a physician to prescribe a drug for any purpose, so long as the drug is approved by the FDA for some purpose. And there is nothing wrong with a manufacturer making huge profits from sales of drugs for these so-called unapproved or off-label uses. The FDA's only restriction is that the manufacturer cannot promote the product for unapproved uses, for example, in advertising to the public or to physicians at medical meetings. But even this restriction is unraveling. The FDA Modernization Act of 1997 now allows manufacturers to distribute literature on an unapproved use if the manufacturer is conducting studies that might lead to eventual FDA approval for the use, and federal courts have held that FDA limitations on disseminating information about unapproved uses must be narrowly tailored to avoid violating the sponsor's constitutionally protected right of free speech.[24]

That's why Pfizer earns more than $1.5 billion a year on sales of Viagra, a drug that, although indicated only for the medical condition in males known as erectile dysfunction, is used as a sexual enhancer by millions of men with normal sexual functioning, and even by many women.

The experience with human growth hormone (HGH) illustrates how the unapproved use phenomenon might work in connection with genetic enhancement. The drug is approved by the FDA to treat children with growth hormone deficiency. But pediatric endocrinologists began prescribing it for children with short stature, regardless of whether this was attributable to a hormone deficiency. There are even reports that some parents seek the drug for children who are naturally tall, in the hopes that the kids will gain an extra few inches and be more likely to become successful professional basketball players.[25]

The result of the regulatory lacuna created by the FDA's lack of control over unapproved uses is that without having to obtain the agency's approval for enhancement claims, manufacturers of enhancement products will have no need to generate safety and efficacy data for these uses. Some indication of how safe and effective the product is for enhancement use may be gleaned from the data on its medical uses that the manufacturer submits to the FDA, but since efficacy and safety can vary widely for different uses in diverse populations, people who want to use the product for enhancement purposes by and large will end up serving as their own guinea pigs, at least until any serious side effects show up after long-term enhancement use by large numbers of people. By then, much harm may have occurred, and lives may be lost.

The solution is to fill in these regulatory gaps—to give the FDA, or some equivalent government agency, the power to regulate both the practice of medicine and off-label prescribing, at least when they involve enhancements. This is something Congress could do if it had the political will to take on the American Medical Association and other similar organizations, which remain steadfastly opposed to almost all government restrictions on physician autonomy. One of the few exceptions is physician-assisted suicide, which citizens of Oregon have twice voted to permit in their state but which the AMA believes should be outlawed as inappropriate behavior by physicians.[26] Interestingly, the Oregon assisted-suicide law furnishes an illustration of one mechanism that the government might employ to regulate off-label uses: Physicians who prescribe a lethal dose of drugs for their patients under the Oregon law must indicate on the prescription itself that this is their intent. A similar requirement that physicians disclose the enhancement purpose of a prescription would give the government an idea of how much enhancement use was taking place and by whom, affording a chance to monitor the health status of the consumers involved. The FDA could even be given the authority to require manufacturers whose products were being used for enhancement purposes to conduct safety and efficacy studies on the enhancement uses, even if the manufacturer had not sought government approval for enhancement claims.

But the prospect that manufacturers might test products for enhancement uses—either because they chose to obtain FDA approval for enhancement claims or because, under new legislative authority, the agency forced them to conduct testing because of widespread de facto enhancement use—raises a further serious ethical problem. Under what circumstances, if any, would it be appropriate to conduct human tests of a product, say a drug, to ascertain its enhancement rather than therapeutic value? Generally speaking, it is ethical to test a product in humans only if its benefits are expected to outweigh its risks. But recall the earlier discussion of the subjective nature of enhancement benefits. How would we—or the government—determine whether a genetic enhancement was likely to produce more good than harm? How would we measure the benefit? And how would we offset this benefit against the threat of serious adverse health effects, including some that might not show up for many years?

This brings us back to the fundamental question with which this chapter began: How much should the government interfere with individual choice? If people want to enroll in a genetic enhancement experiment, why shouldn't they be allowed to? According to this approach, the appropriate role for the government might be merely making sure that the individual decision makers were given all the information relevant to making an informed choice. To this end, the FDA might require the manufacturer to report known or suspected hazards and predict success rates based on animal studies, perhaps using government-mandated reporting standards. This is the approach employed, for example, with in vitro fertilization clinics, which are unregulated by the FDA but which are required under federal law to report in standard fashion their rates of success in producing live births.

Even under a stronger model of government regulation to ensure safety and efficacy, the FDA might leave people relatively free to choose genetic enhancements that presented a low risk of adverse effects. There might be a class of enhancement products that were so safe that they could be sold over the counter without a prescription, like common cold and pain remedies. And one other type of enhancement intervention may be particularly appropriate for a hands-off government attitude: the passive forms of enhancement described in Chapter 3, like selecting among fertilized embryos the ones to implant that have the "best" genetic profiles, or aborting fetuses when the parents were disappointed with the results of in utero genetic testing. The law now leaves parents free to decide which embryos to implant in the course of IVF and whether or not to abort a fetus in the early stages of pregnancy based on genetic tests designed to reveal diseases or susceptibilities for illness. A strong argument can be made that parents also should be left free to make similar decisions after genetic testing for enhancement purposes.

But the mention of embryos and fetuses suggests another set of complications. Embryos and fetuses cannot make choices. Neither can young children. Yet a good deal of genetic enhancement is likely to be aimed at children, who may not be capable of making informed decisions about whether or not to participate in enhancement experiments. Would it ever be permissible for a manufacturer to conduct enhancement experiments on kids? Presently, there are all sorts of special legal protections for children who seek to be enrolled in medical experiments. The gene therapy experiment described in Chapter 3 in which Jesse Gelsinger died furnishes an illustration. Based on bioethicist Arthur Caplan's argument about the importance of being able to make an informed choice, the experiment was carried out in relatively healthy adult volunteers like Gelsinger rather than in the severely ill newborns who were the eventual targets of the therapy if it proved successful. One court in Maryland has gone so far as to rule that children cannot be enrolled in any medical experiment that does not offer the prospect of direct medical benefit to them, even if the risks are minimal.[27] The law might take an even dimmer view of experiments whose objective was to enhance children rather than to benefit them medically.

The prospect of genetic enhancement in children also raises a different sort of problem. We have been considering two models of enhancement decision making: one in which the government decides whether an enhancement is safe and effective enough to be marketed, and another in which the decision is left up to fully informed individuals acting on their own volition. But children who are given enhancements or who are enrolled in enhancement experiments may not be acting voluntarily. Enhancements may be imposed on them by their parents without the child's knowledge or consent. Should this be permissible? This leads to a broader question: To what extent should people be able to force others to use enhancements? What if employers, for example, insisted that workers use genetic enhancements to increase their productivity? In short, how do we make sure that enhancement decisions are autonomous?

6

Autonomy

Among the fundamental tenets of modern biomedical ethics, none is more basic than the right of self-determination. Physicians must obtain their patients' permission before proceeding to treat them, and acquiesce to a refusal even if it means the patient's death.

Despite its centrality, self-determination in medical decision making is a relatively recent development. Until the early 1970s, physicians rarely left matters up to patients. The relationship between patients and physicians was paternalistic, so much so that it was called the "physician-patient relationship," as if the physician were the more important party. Hippocrates even discouraged physicians from telling patients what was wrong with them, advising doctors to perform their duties "calmly and adroitly, concealing most things from the patient while you are attending to him":

> Give necessary orders with cheerfulness and sincerity, turning his attention away from what is being done to him; sometimes reprove sharply and emphatically and sometimes comfort with solicitude and attention, revealing nothing of the patient's future or present condition.[1]

It was not until the patients' rights movement in medicine in the late 1960s, which arguably grew out of the general distrust of authority characteristic of those times, that courts and bioethicists began to give prominence to the principle of patient autonomy. The law began to acknowledge that physicians had a duty to obtain their patients' informed consent to treatment, and patients gained the right to refuse to go along with what the doctor recommended, even at the cost of their lives.

The case of *Lauro* v. *Travelers Insurance Company* illustrates the traditional view of the law.[2] Mrs. Lauro found a lump in her breast. She underwent a biopsy, in which she was placed under general anesthesia, an incision was made, and a sample of breast tissue removed and sent to a pathologist, Dr. Nix. Dr. Nix performed a quick microscopic analysis of the tissue and concluded that it was cancerous. Mrs. Lauro's right breast was removed while she was still under anesthesia. Later, after a more leisurely and careful analysis, Dr. Nix concluded that the lump had been a rare type of benign tumor that was difficult to distinguish from cancer using the rapid technique she had employed. Mrs. Lauro sued Dr. Nix for malpractice and ultimately lost.

The debate between the judges on the panel that heard the case is what is instructive. A majority felt that Dr. Nix had acted properly even though she had been mistaken, while a dissenting judge felt that Dr. Nix should not have rendered an unequivocal diagnosis of cancer based on the rapid analytic technique she had used. The majority pointed to expert testimony about the risks of delay in case the tumor was in fact malignant, and the dangers of having to reanesthetize Mrs. Lauro if a subsequent mastectomy turned out to be necessary. The dissent argued that, based on testimony from other experts, a delay of a few days while the more thorough study of the tissue was conducted would not have created a significant risk that the cancer would spread, and that the hazards to Mrs. Lauro from being anesthetized a second time were outweighed by the benefit of avoiding the unnecessary loss of her breast.

What is striking about the case is that in neither the majority's or the dissent's discussion is there any mention whatsoever of what Mrs. Lauro herself preferred. Evidently she wasn't asked ahead of time which set of risks she wished to accept; her physicians had acted on their own initiative, and the outcome of the case hinged on the judges' estimation of what *they*, rather than the patient, would have chosen to do under the circumstances.

At the same time these judges in New Jersey were debating their preferences regarding Mrs. Lauro's predicament, another case decided in the District of Columbia showed that the law was beginning to change dramatically. The D.C. case, *Canterbury* v. *Spence*, involved a patient who had become paralyzed after a laminectomy—an operation to remove a protruding portion of a slipped disk that was causing him severe pain.[3] The plaintiff alleged that the surgeon had not warned him of the risk of paralysis, and therefore had proceeded without his consent. Judge Higgenbotham laid down the principles that define the modern relationship between physicians and their patients: the physician must inform the patient of what is wrong; of the alternative ways of proceeding; and of the material risks and benefits of each alternative. If the patient can prove that, had he been informed of a risk that later materialized, he would not have gone forward with the treatment, he can recover compensatory damages.

If this modern approach had been applied in the *Lauro* case, Dr. Nix would have been liable unless she could prove that the risk of a mistaken diagnosis was so remote that, even if Mrs. Lauro had been told about it, it would have been immaterial to her decision about how to proceed. The chief decision maker would have been Mrs. Lauro, not the physicians and certainly not a panel of appellate judges.

The doctrine of informed consent also plays a major role in modern medical research. Under rules adopted after the details of Nazi medical experiments on concentration camp inmates became public, investigators cannot conduct research on human subjects without their consent. The researchers must explain the risks and benefits of their experiments to potential subjects before enrolling them in clinical trials, even though the knowledge gained from the research may be vital to saving lives. Special protections kick in for vulnerable subjects such as children, prisoners who may view cooperation with researchers as a ticket to favorable treatment by the authorities, and individuals with mental impairments.

Yet the doctrine of informed consent is full of flaws. There is mounting evidence that most patients cannot understand or accurately process the information given them by their physicians.[4] Since most patients simply acquiesce with whatever the clinician recommends, all that the information may accomplish is to frighten patients with visions of horrible side effects that never occur. In the research context, the potential risks and benefits are often highly speculative, since they are extrapolated from laboratory and animal data that may not apply to humans. In addition, the Institutional Review Boards charged with making sure that researchers have obtained informed consent from their subjects do not always do their jobs properly. One reason is that they are understaffed and overwhelmed by the amount of work they are expected to perform. Another problem is that most of the IRB members are employed by the same institution as the researchers proposing the studies, and the reviewers therefore have an interest in the study going forward to secure research funds and glory for their institution. Another conflict of interest arises when physicians attempt to enroll their patients as subjects in their own experiments. This blurs the line between the physicians' roles as researcher and therapy-giver and can confuse patients into thinking that the experimental intervention actually has proven therapeutic value. After all, why would your physician enroll you in an experiment that wasn't going to do you any good? In this and other contexts, health care professionals find themselves torn between their own self-interest—for example, in obtaining the financial and professional rewards of high-profile biomedical research—and their obligation to act only in their patients' best interests. As noted in Chapter 3, financial conflicts of interest are implicated by some in the death of Jesse Gelsinger.

But there is a more fundamental problem with the notion of autonomy as applied to medical treatment and research. The entire concept of self-determination, connoting an exercise of free will, often has questionable applicability in the dire situations in which patients and research subjects find themselves. For people who are seriously ill, sometimes at death's door, the treatment being offered, or the option of becoming part of a research investigation and receiving an experimental therapy, may represent their only hope. In these circumstances, are their decisions voluntary, or does their condition serve as a source of duress? Do these individuals really have a choice?

These same concerns arise in the case of genetic enhancements. Individuals contemplating whether or not to employ genetic enhancements may not receive complete, truthful risk information from physicians and others who are eager to make a profit or to secure subjects for lucrative or professionally rewarding research. As noted in Chapter 3, moreover, persons deciding whether or not to use enhancements may feel that they are under considerable duress. The more effective the enhancements, the greater the pressure to use them. Fearful of being unable to compete successfully for scarce societal resources, people may feel they have no option but to enhance themselves or their children.

A similar phenomenon occurs in sports, where athletes, often pressured by their trainers and coaches, feel bound to use performance-enhancing drugs such as anabolic steroids. As Thomas Murray notes, if the drugs provide a performance advantage, and any athlete uses them, then all athletes must use them or give up hope of winning.[5] One genetic enhancement already available is synthetic erythropoeitin (EPO)—"genetic" because it is manufactured by recombinant DNA techniques—which athletes use to increase the number of red blood cells in their bodies so that more oxygen is transported to their muscles, thereby improving physical performance.

One of the chief criticisms of performance-enhancing drugs in sports is their health risks. Steroids, for example, can cause heart and liver problems.[6] Athletes who have no choice but to use these products to remain competitive are being coerced into grave physical peril. This is one key reason that sports organizations such as the International Olympic Committee ban the use of these products, and go to great lengths to implement effective programs to detect and punish violators.

If genetic enhancements also were fraught with health risks, their use might raise similar concerns. Acting out of a desire to protect people who felt they had no choice but to incur these risks, we might prohibit the use of enhancements and disqualify those who ignored the ban from enjoying the fruits of their success.

But how concerned should we be about the pressure to use genetic enhancements, or for that matter, performance-enhancing drugs in sports or risky experimental treatments? Isn't "pressure" just another way of saying that these interventions can provide substantial benefits? People who choose to use genetic enhancements out of a desire to succeed, like athletes and patients, still have a choice. Patients can refuse treatment, although this might harm them or bring on their deaths, and athletes can refuse to use performance-enhancing drugs, although perhaps at the cost of their athletic careers. So, too, individuals ought to be able to decide for themselves whether to incur the health risks of genetic enhancements in return for the expected benefits.

Indeed, we allow people to take risks all the time: the risk of flying or driving cars, the risk of participating in sports like skiing and mountain climbing. We might hesitate if this were the only way these people could obtain the basic necessities of life, like food and shelter, but we seem quite willing to permit people to assume enormous risks for the sake of pleasure, fame, or fortune, so long as they do not harm others in the process. At the same time that the Olympics cracks down on athletes who use performance-enhancing drugs, for example, it permits them to subject themselves to brutal physical training programs, which cause serious physical and mental injuries.

In fact, it may be unfair to prevent people from using genetic enhancements. Performance-enhancing drugs in sports may make up for disadvantages like being unable to afford a top trainer or being born with a slight frame. Drug use may be the only way that some persons can engage in certain sports, let alone become serious competitors. Similarly, genetic enhancements can be used to level the playing field, making up for mental or physical disadvantages and other accidents of birth and fate. The consequences of denying people the option of using these products, even at serious risk to their health, may be to disenfranchise them. If we truly were interested in maximizing individual autonomy, perhaps we ought to let people decide for themselves whether or not to use genetic enhancements, even though they felt they had no choice but to do so in order to be successful, and even though enhancements caused serious health hazards.

Furthermore, the discussion all along has assumed that genetic enhancements, like some performance-enhancing drugs in sports, are dangerous to the user. But they may not be. If enhancements turn out to be relatively safe, is there a problem if people feel pressured to use them in order to be successful?

Yes, there is. The problem is that enhancements are self-defeating. If everybody can get hold of them, and they function effectively for everyone, then they serve no purpose. What good is a performance-enhancing product that adds five miles an hour to a sprinter's speed, or enables ice skaters to make an extra revolution in their jumps, or adds twenty IQ points to a person's intelligence, if it does so for everybody and everybody uses them? Everyone would be competing just as they did before, only at faster speeds or with more jumps or brainpower. Enhancements would provide no competitive advantage, but everyone would have to use them or fall by the roadside. Even if they entailed no serious health risks, enhancements would be a waste of resources. It would be appropriate therefore to ban their use under these circumstances; the only losers would be the providers of the enhancements, who would not be able to get rich at the competitors' expense, and perhaps spectators who would not be able to watch as exciting an event. And if the enhancements did create health hazards to the users, the rationale for banning their use would be all the more compelling.

Of course, the situation would become very different if not everyone could get hold of the enhancements. Then there would indeed be advantages for those with access to them, advantages that quite possibly were decisive. But this arguably would be an unfair advantage. Therefore, we might deny people the option of using an enhancement, not only because we were concerned about their autonomy, but to promote fairness.

For example, what if genetic enhancement were not available to everyone, and an employer refused to hire anyone who was not genetically enhanced, or required employees to take enhancement drugs? This would create a powerful form of unfair coercion, assuming jobs were scarce or the employment in question was highly desirable or remunerative. Should this sort of employer duress be permitted?

Under current law, the answer would depend on how the Americans with Disabilities Act was interpreted to apply to genetic enhancements. The act permits employers to test employees for conditions or traits that are "job-related." If the conditions being tested for also qualify as disabilities under the act, the employer must attempt to reasonably accommodate the employees' disabilities. Employers can take adverse action against the employees with disabilities—such as refusing to hire them, demoting them, or firing them—only if the employees remain unqualified for the job despite the employer's accommodations.

Translated into the realm of genetic enhancement, the law would only apply to employees who were not enhanced if "not being enhanced" were regarded as a disability. The term "disability" is defined in the law to mean a condition that substantially interferes with a major life activity, including working. Although when Congress passed the law, it clearly did not have the possibility of genetic enhancements in mind, the condition of not being enhanced might well be regarded as a disability if it prevented the unenhanced from working in desirable jobs. Under this interpretation, the law would require an employer to hire people who were not enhanced if they could qualify for the job with "reasonable accommodation." The accommodation might mean that the employer would have to accept the fact that unenhanced employees work less efficiently than enhanced employees. So factory management might have to slow down the production line, or tolerate a larger number of mistakes. By forcing the employer to accept these reductions in productivity or quality, the law eliminates the employer's incentive for preferring enhanced workers in the first place, which is exactly what the anti-discrimination law is designed to accomplish.

But the analysis might come out differently if the employer's motive were not merely to increase productivity. Consider the professional rescuers in the Introduction. Would their employer be forbidden from insisting that they take enhancements while on the job, or from hiring only individuals whose enhanced traits made them exceptionally good at rescuing? These employer preferences would still be coercive or unfair, but these criticisms might be outweighed by the public benefit provided by having really good rescuers. What if the employer were an airline, and wanted all pilots to be enhanced so they had better vision or other superior flying skills? What if the employer were a nuclear power plant and insisted that all employees be enhanced so that they were able to act coolly in an emergency? In all these instances, the employers would be motivated by the desire to prevent accidents or to benefit third parties like rescue victims, rather than simply to increase profits. Therefore, we might allow the employers to impose these conditions on employees, even though it would rob those who wanted the jobs of their autonomy to avoid becoming enhanced, and although it would discriminate against persons who could not become enhanced even though they wanted to.

Would it ever be appropriate for the government to require people to become enhanced? If the purpose behind doing so was to create a better class of citizens—persons with a better genetic endowment—this would be a high-tech version of old-fashioned eugenics. Presumably the lessons of the early part of the twentieth century, and especially of the Nazis, would still be fresh enough in people's minds that the public would not stand for this. But suppose we were at war, and the government wanted to create a group of enhanced soldiery? In time of national emergency, the Constitution gives the government virtually unlimited powers to ensure the nation's survival. Just as it can require soldiers to go through training and into combat, whether or not the soldiers want to, the government arguably could require them to become enhanced in order to make them more effective or better at performing specialized tasks.

Yet even here, there may be limits to what the government can require citizens to do. The government has been roundly criticized for subjecting soldiers during the Cold War to drug and radiation experiments without their consent and, in some cases, without their knowledge. With an all-voluntary military rather than draftees, the government might feel it would have greater latitude to require soldiers to become enhanced, since this might be made one of the conditions for enlisting. But even then, the government might be criticized if it enhanced soldiers without their knowledge.

More difficult issues of lack of autonomy arise when genetic enhancement is performed on embryos, fetuses, or children. Certainly embryos and fetuses cannot agree to become enhanced, and although children, depending on their age and sophistication, may be able to understand something about what enhancement entails and to express their personal preferences, they are incapable of providing full-fledged informed consent.

One criticism that is sometimes voiced against genetically enhancing children is that it would deprive them of their right to a natural genetic destiny. Dinesh D'Souza states, for example, that "the good parent will respect the child's right to follow his own path":

> It is unconvincing to argue, as some techno-utopians do, that giving a child a heightened genetic capacity for music or athletics or intelligence is no different from giving a child piano, swimming, or math lessons. In fact, there is a big difference. It is one thing to take a person's given nature and given capacity, and seek to develop it, and quite another to shape that person's nature in accordance with one's will.[7]

With passive forms of genetic enhancement, embryos and fetuses clearly do lose their futures, genetic and otherwise, since the deselected embryos are discarded and the "suboptimal" fetuses are aborted. But current law gives parents, and biological mothers in particular, the legal right to impose these fates for whatever reasons they choose, so long as they act early enough in the gestational process. Some people may reject IVF and non-therapeutic abortion and urge their views on others. But unless the law is changed—and it is not clear that it should be—these techniques, and the passive genetic enhancement they make possible, will continue to be constitutionally protected, and the government will be permitted to interfere with them only to protect the health of the mother, regardless of the lethal consequences for the embryos and fetuses.

In the case of children, the criticism that they lose their genetic future is even more unpersuasive. The "natural genetic destiny" of a child who is born with genetic illnesses or impairments is to suffer or to die, but parents clearly have the right, and arguably a moral and legal obligation, to alter that destiny by intervening medically if they can do so safely and effectively. No valid ethical distinction can be drawn between an undesirable and a desirable natural genetic endowment. Neither is deserved. Neither is etched in moral stone. If parents ethically can prevent their children from enduring the pain and suffering of genetic ailments, they ethically can permit them to enjoy the pleasures and rewards of genetic advantages. Moreover, parents clearly can act to benefit their children without the child's assent, and even over their sometimes passionate objections. Parents do all sorts of things to give their children advantages in life, like making them do their homework or practice a musical instrument, that the children resent. Of course there are limits to what parents can do. The woman mentioned in Chapter 4 who hired a hit man to kill a cheerleader's mother so her daughter could take her place on the squad went too far, as perhaps did the parents who fed their toddler okra prior to her preschool interview. But the concern in these cases is with the physical or psychological harm that these practices may cause, not with the child's loss of an open future. So long as a genetic enhancement is safe and effective, there can be little objection on grounds of autonomy to the practice of parents enhancing their children.

But concern for the physical or mental well-being of the child clearly *is* a valid reason for limiting parental autonomy. As noted in the previous chapter, genetic enhancement may not be safe, and parents may not be free to impose its risks on their children or to appear to allow the children to accept the risks for themselves. If the hazards posed by genetic enhancement are significant, or are not obviously outweighed by the benefits, a child's immaturity and lack of sophistication become significant concerns, and parents should not be allowed to pre-empt the child's autonomy to decide whether or not to accept these risks for itself. Risky enhancement instead should wait until children become mature enough to make their own decisions about how to weigh the risks and benefits.

At what age children reach that stage, however, is a matter of dispute. Some youngsters clearly are capable of making informed decisions before they reach the legal age of majority. The law in most states recognizes the "mature minor" doctrine, which allows adolescents to make medical treatment and other important decisions for themselves if they are capable of understanding and balancing the risks and benefits. Presumably this is the principle that sustains cosmetic surgery, such as nose or breast jobs, when it is performed on teenagers, although one wonders about the wisdom of some parents in Southern California who are reported to give their daughters breast augmentation as high school graduation presents.[8] If these same approaches were applied to genetic enhancement, then minors who were old enough to make mature decisions would be permitted to go along with parental efforts to enhance them, and might even be able to obtain enhancements on their own without parental permission.

Experimenting on children is a different matter, however. The law bends over backward to allow children to avoid the risks involved in being experimental subjects. Federal regulations, for example, require researchers to obtain the assent of even very young children before enrolling them in medical experiments.[9] By the same token, children who are capable of expressing their wishes must be given the opportunity to refuse to be part of experimental attempts at genetic enhancement.

7

Authenticity

If individuals knowingly and voluntarily agreed to be genetically enhanced, or if the benefits so clearly outweighed the risks that we permitted children and others to be enhanced despite their inability to consent, we would face the question of how to value their accomplishments. Should a race won by a genetically enhanced athlete count the same as a race won by dint of years of training and self-sacrifice? Is a college degree obtained with the aid of genetic enhancements a legitimate reflection of individual achievement? Would the fact that an artist had been genetically enhanced diminish the beauty of a painting or sculpture? In short, how valid are accomplishments made with the aid of genetic enhancement? Are they true achievements? Are they authentic?

This problem has arisen historically with athletes who use drugs to enhance their performance. The organizations that sponsor the events they compete in ban the use of these drugs and go to great lengths to detect and punish violators. The most prominent of these organizations is the International Olympic Committee (IOC). The IOC first became concerned about performance enhancement—which it calls "doping"—in 1960, and in 1961 it created a Medical Commission to deal with it. The basic anti-doping principles of the IOC were laid down in 1967: (1) "protection of athletes' health"; (2) "defense of sports ethics"; and (3) "equality for all participants at the moment of competition."[1]

The first drug tests were performed at the 1968 Winter Olympics in Grenoble, France. A subcommission on "Doping and Biochemistry of Sport" was established in 1981 to identify banned substances and practices, to promulgate approved testing methods, and to certify testing laboratories. Beginning in the early 1990s, the IOC broadened its concern to include the use of doping during training outside of competition. In 1999, an independent body called the World Anti-Doping Agency was established in Lausanne, Switzerland, to coordinate the worldwide anti-doping effort.

At the 2000 Summer Games in Sydney, Australia, a total of 2,846 anti-doping tests were performed. Forty-one athletes were disqualified before they even reached Australia, nine after they arrived but before they could compete.[2] Three Bulgarian weightlifters forfeited medals after they tested positive for a drug that can mask the use of steroids. A Latvian rower and a Russian runner were stripped of medals for steroid use, and in the most controversial case, a sixteen-year-old Romanian gymnast lost her gold medal after testing positive for an over-the-counter medicine that contained a banned stimulant, which her team doctor apparently had given her to treat a cold.

One drug that the IOC tests for is erythropoeitin (EPO), which stimulates the production of red blood cells. Since EPO is made using recombinant DNA, it qualifies both as doping and as a genetic enhancement. Olympic officials detected approximately 100 uses of EPO during the 2002 Salt Lake Winter Games.[3] In short, the IOC already is testing athletes for the use of a banned genetic enhancement substance.

As the IOC's anti-doping principles state, a major concern is the negative effect of performance-enhancing drugs on the athletes' health. Many of these substances can cause serious health problems, even death. But some of these drugs, such as the over-the-counter medicine given to the Romanian gymnast, are quite safe. And others, although they may present a risk of adverse effects, are no more hazardous than some of the physical and emotional abuse that athletes subject themselves to during training. Why are these substances banned as well?

The answer lies in the IOC's other two anti-doping principles, "defense of sports ethics" and "equality for all participants at the moment of competition." Even if the substances are not harmful to athletes, the IOC is saying, their use is unethical. They confound the integrity of sport. Athletes who use them are cheating. They don't deserve to win. Their victories are inauthentic.

The same charges can be lodged against users of genetic enhancements. The advantages conferred by enhancement, it can be argued, are not merited. An athlete who places first by virtue of using genetic enhancement no more deserves a medal than an athlete who uses steroids or EPO.

But let's take a closer look at the IOC's second and third rationales. The third—"equality for all participants at the moment of competition," gets back to the question of whether enhancements are readily available to everyone. If they are, and if they are relatively safe, then they don't make competitors unequal, since all athletes can have access to enhancement advantages. The previous chapter noted one problem, though: If genetic enhancements merely ratchet up everyone's abilities by a certain fixed amount—like an extra twenty IQ points—then enhancements that were available to everyone would be pointless, since they would confer no competitive advantage. On the other hand, any competitor who failed to take them would be at a disadvantage. This seems to be a valid reason for banning their use, not because it makes competitors unequal, but because it would be stupid.

Actually, genetic enhancement might promote equality among athletes. As pointed out in the previous chapter, athletes are far from equal at the moment of competition. They are not "naturally equal," since they were born with different characteristics such as stature and coordination. And they can't all hire the best trainers and coaches or secure the best equipment. So one way to increase equality would be to allow genetic enhancement to be used by competitors who were disadvantaged in other respects for reasons beyond their control. Furthermore, instead of conferring a fixed amount of advantage, genetic enhancement may enable its users to attain a certain common level of ability. In other words, it might be able to give everyone the ability to bench press 300 pounds. In that case, again, its use in sports would increase rather than decrease equality. But then who would want to watch a race where, barring bad luck, everyone arrived at the finish line at the same time?

In the case of widely available, relatively safe genetic enhancements, the second of the IOC's rationales—"defense of sports ethics"—may be more persuasive. The notion is that it is unethical to use performance-enhancing drugs, and by extension, genetic enhancement. One reason is that they are shortcuts that enable an athlete to succeed without much effort, to win without hours of practice and years of dedication and hard work. But why is hard work so important? It hardly seems worthwhile in itself. Is it just an artifact left over from our Puritan ancestors, a hang-up based on a particular conception of "God as taskmaster"? Shouldn't cleverness, for example, be just as admirable?

There is a story they tell about the exams that English schoolchildren take at the age of eight or so to determine whether they should be tracked in the direction of a university or a technical education. At one time, one part of the exam was an essay, and it seems that for years, the question was always the same: "Write an essay on Sir Frances Drake." Since everyone knew what the question was going to be, the exam turned out to test merely the ability to memorize the essay the children had practiced with their tutors, plus penmanship.

But one year they changed the question. Without warning, the children were asked to write an essay on dogs. One child, the story goes, began his essay with: "There are many different types of dogs: terriers, collies, Alsatians . . . and sea dogs. Sir Frances Drake was a sea dog. . . ." Then the child went on to copy down his essay on Sir Frances Drake.

Are we horrified by this account, or amused? Wasn't this child's cleverness and ingenuity admirable? Should he get a lower mark than the child who kept on writing about canines, or a higher one?

There might be a legitimate concern, of course, with the authenticity of test results achieved with the use of somatic genetic enhancements the effects of which were transient. These enhancements would wear off, and there would be no guarantee that the individual would continue to take them in the future. Therefore, performance on exams designed to predict future performance, like SATs or law school entrance tests, might be invalidated by the use of transient somatic genetic enhancements. But this concern would not arise in the case of a permanent enhancement, like pre-implantation gene transfer or deletion.

Aside from that, though, there doesn't seem to be any *a priori* reason why the use of enhancements, assuming they are safe and readily available, would be unethical. It is merely that they are contrary to the rules of the competition. In other words, they are simply not how the game is played.

But it is hard to find a reason why the rules prohibit them, other than tradition. There is no *a priori* reason why competitions couldn't allow the use of enhancements, just as there's no inherent reason that a baseball game ordinarily runs nine innings. Those are the rules, but the rules can be changed. There have been numerous examples of rules that were changed to permit athletes to use techniques or equipment that otherwise would be considered cheating. Until the 1960s, for instance, the poles used by pole-vaulters were made of wood. Then someone invented fiberglass poles, which enabled athletes to clear a much greater height. Instead of banning the use of the fiberglass poles, the rules were changed to make them legal. The rules in other sports have been changed as well. Tennis rackets have grown larger. Football permits referees to review plays on videotape.

If some athletes still think it's wrong to use enhancements, then we can hold twin athletic events: an enhanced and unenhanced 100-meter sprint, for example. Athletes could choose which contest to compete in. This has already been done in the sport of power-lifting, where there are "open" and "natural" events.[4] Harking back to the previous chapter, giving athletes a choice would seem to increase their autonomy. They could even be permitted to decide for themselves whether or not to use enhancements that were hazardous to their health, like the power-lifters who opt for the "open" events and take steroids. So long as the athletes had the option of competing without enhancements, there would be less reason to be concerned that they would feel coerced into making that choice.

For the same reasons, authenticity might not be a good reason to prevent people from using genetic enhancements. So long as enhancements didn't confound predictive testing, and so long as everyone understood that enhancements were permitted, achievements could still count as achievements. What would change would be the nature of the competition: Instead of who ran the fastest, the objective would be to see who ran the fastest with genetic enhancement.

But we must remember the assumptions that we began with. One of them is that genetic enhancement is safe, which it may not be. As noted in the previous chapter, we may feel justified in restricting the use, or at least the experimental use, of dangerous genetic enhancements, especially to prevent them from being forced on individuals who are not in a position to decide for themselves whether to accept the risks. The other assumption is that genetic enhancement is readily available to everyone who wants it. But is that likely? Will supplies be plentiful, or scarce? Will it be affordable, or prohibitively expensive? What if only some people have access to it?

8

Access

The more effective genetic enhancement proves to be, the more it will be in demand. If enhancement turns out to significantly enhance physical attributes like beauty, strength, and coordination, it will be sought by all those who value those attributes or depend on them for their livelihood. People who survive by their memory or their wits or who love the life of the mind will seek enhancements that substantially improve mental functioning. People will want enhancement capabilities to help them accomplish their life's goals. They will feel pressured to enhance themselves to stay ahead, or at least abreast, of the next fellow. They will feel compelled to provide genetic enhancement to their children, so that their children won't be left behind. And if genetic enhancement reaches its full potential, people will reach out with all their might to grab the greatest brass ring of them all—the chance to triumph over aging and ultimately, perhaps, defeat natural death itself.

How easy will it be to satisfy this demand? New technologies sometimes are in short supply because of technical production difficulties. When penicillin was first discovered in the 1930s, it was produced by allowing natural mold to form.[1] At the beginning of World War II, the supply was so limited that it had to be rationed by the military. Interestingly, the soldiers who received it were not ones suffering from battlefield wounds, but soldiers with venereal disease; they were able to return to the fighting immediately, while the wounded required time to recover.[2] Only when mass production methods were devised was there an adequate supply of the wonder drug to treat all who might benefit. It isn't always new technologies that face production limits either. Recall human growth hormone (HGH), which, as described in Chapter 3, was scarce because it had to be harvested from the pituitary glands of cadavers.

Genetic enhancement may face similar supply restrictions, although for different reasons. Embryo selection for enhancement purposes—in which embryos are created in the test tube, tested for non-disease (as well as presumably disease) characteristics, and only the "best" implanted in the womb—will be limited by the availability of viable eggs and sperm and by the natural frequency of their genetic characteristics. Abortion for enhancement purposes, in which fetuses are aborted if they do not score well enough in prenatal genetic testing for desired traits, is limited by the number of a person's possible pregnancies. More technologically sophisticated forms of genetic enhancement, involving gene insertion and deletion, may be in short supply because of the dearth of physicians and clinics capable of performing the techniques successfully.

But the shortage of competent professionals will be short-lived. The enormous demand for enhancement, fueled by enticing descriptions of the genetically rich and famous, will drive numbers of health care professionals into the lucrative enhancement business. Increasingly frustrated by managed care, with its limitations on fees and interference with clinical judgment, physicians will turn to enhancement as a new source of revenue, the way many of them have embraced so-called alternative and complementary medicine. Hospitals and clinics will view enhancement "patients" as a way to fill their beds and ambulatory surgery suites, leading them to establish new enhancement wings and centers and to recruit additional staff.

Moreover, recombinant DNA will create virtually unlimited supplies of somatic enhancement drugs and similar products. Recombinant DNA manufacturing, it will be recalled, enabled supplies of human growth hormone to become so plentiful that parents were able to procure it for children who weren't hormone-deficient. Difficult tasks lay ahead: identifying the regions of the genome that code for non-disease traits, understanding the interactions between these genes and between the genes and the environment, and developing accurate genetic tests for these traits. But once these tasks are accomplished, which will be only a matter of time, there will be no significant technical impediments to manufacturing large quantities of enhancement drugs.

All this isn't to say that genetic enhancement will be easy to come by. A growing number of health care professionals may be capable of providing enhancement services, and drug companies may discover and produce large amounts of enhancement drugs, but they are likely to be very expensive. Human growth hormone therapy for a twenty-kilogram child costs in the neighborhood of $14,000 per year.[3] Genetic testing costs between $300 and $3,000 per test. In vitro fertilization (IVF) alone now costs an average of about $37,000 per live birth, not counting the costs of genetic testing or of genetically manipulating the embryos.[4]

One of the reasons for the high cost of enhancement products is the U.S. patent system and its impact on drug research and development and on government licensing.

The Human Genome Project—the genome sequencing program funded and administered by the government—is committed to placing all of its discoveries in the public realm, free for anyone to consult to make further discoveries. But Celera and other private companies patent their research findings, and refuse to allow others to use them or permit them to be used only upon the payment of substantial royalties, which jacks up the price of any eventual products or services.

Toward the beginning of the current revolution in human genetics, a debate arose over who would own the resulting discoveries. There was general agreement that traditional inventions like drugs could be patented even though they were developed with the knowledge gained from genetic science. But many people objected to the notion that you could patent a living organism, such as a bacterium that had been genetically engineered through recombinant DNA to produce a desired substance, like human growth hormone. In 1980, in a case involving bacteria that had been engineered to consume oil spills, the U.S. Supreme Court ruled that a living organism could be patented so long as it was a new entity that did not occur in nature.[5] Later, courts upheld patents taken out on actual human nucleotide sequences, like the gene for EPO, which could be inserted into the recombinant bacteria to produce quantities of EPO.[6] The judges reasoned that, since the researchers seeking the patent had isolated the DNA sequences that coded for the 165 amino acids that produced the EPO protein, they deserved the twenty-year monopoly conferred by the patent laws. Currently, thousands of patent applications are pending merely for gene fragments—stretches of DNA sequences identified by sequencing machines—even though the researchers know little or nothing about their function.[7]

The effects of permitting private ownership of the human genome are plain to see. For example, one company, Myriad Genetics, has patented a gene associated with breast cancer. Anyone who wants to use the genetic test that Myriad developed to see if someone is at elevated risk for the disease has to pay Myriad's price, and the company reportedly has used its patent to block research on the gene by others, such as the University of Pennsylvania.[8]

The cost of genetic enhancements, like most products, undoubtedly will decline with time. Production methods will be refined. Testing will be developed that permits a battery of genetic tests for a large number of characteristics to be performed simultaneously. Discoveries eventually lose their patent protection, allowing the marketing of cheaper generic forms of products and services. As more health professionals and facilities begin offering genetic enhancements, competition should force them to drop their prices.

But genetic enhancements will remain unaffordable for many people. In 2001, more than thirty-two million Americans had incomes below the poverty level (set at $9,214 for an individual below the age of sixty-five).[9] The median U.S. household income in 2001 was $42,228. No one below the poverty line, and few people below the median household income level, are going to have disposable income that they can devote to the purchase of genetic enhancements. At best, they may be able to secure some of the cheaper enhancement drugs, enough to perhaps enhance one or two characteristics over a short time. But they will not have the wherewithal to purchase genetic enhancements costing thousands of dollars. Consequently, they will not have access to the more powerful and long-lasting enhancements, the ones that depend on gene infusions or genetic modification for their effectiveness. They will not be able to afford to actively enhance their children. And they will not be able to modify their germ lines.

Aren't we forgetting health insurance, though? Approximately forty million Americans without insurance might not have access to enhancements, but won't insurance go a long way toward providing affordable enhancements for most people, including some of the poorest, who are eligible for Medicaid?

The answer is no. Health insurance will not cover genetic enhancements, any more than it now covers cosmetic surgery. Private health insurance invariably contains an exclusion for services and supplies primarily to improve appearance and for services that are not "medically necessary." Federal Medicare legislation prohibits paying for "items or services . . . which are not reasonable or necessary for the treatment of illness, or to improve the functioning of a malformed body part," and specifically excludes coverage of cosmetic surgery.[10] State Medicaid programs impose similar restrictions.[11]

These exclusions are strictly enforced. An example is a 1995 case from Idaho, where a family sought to force the Idaho Medicaid program to pay for surgery to reduce the size of their son's unusually large ears.[12] Backed by expert testimony from physicians and child development specialists, the parents argued that the otoplasty, as the surgery is called, was not cosmetic, but instead was necessary to prevent the teasing that was damaging the developmentally disabled eight-year-old's self esteem. Even though the boy's ears might have been considered a malformed body part, and hence correctable under Medicaid, the court sided with the state Medicaid program, which maintained that the surgery was excluded as cosmetic.

Some private health insurance plans cover infertility treatments, including in vitro fertilization in some cases, because state law requires them to.[13] As we saw in Chapter 4, in vitro fertilization is a precursor to some forms of genetic enhancement, and some parents might therefore be able to obtain insurance payment for a portion of the costs of enhancing embryos. But insurers would not pay for the costs of the enhancement itself, or for the genetic testing that went along with it, and would attempt to exclude payment for in vitro fertilization when it was performed primarily for enhancement purposes rather than to treat infertility. Furthermore, most people get private health insurance through employment plans that under federal law are exempt from state regulation and therefore not required to comply with state legal mandates concerning coverage of infertility treatments.[14]

Given the demand for genetic enhancement, you might think insurers would market special insurance policies that paid for it, like dental or nursing home insurance or policies that pay if your vacation is canceled or your eyeglasses break. But enhancement insurance won't be sold because there would be no point in buying it. The reason has to do with the fundamental nature of insurance. Insurance is based on the assumption that, on average, the insureds will pay more in premiums than the insurance plan has to pay out in claims. In the case of health insurance, for example, insurers calculate their premiums by estimating the future consumption of health services by their policyholders. In essence, the insurance companies are betting that the premiums from those who don't consume health services, or who consume less, will more than subsidize those who get sick and require major health expenditures. There would be no point in buying genetic enhancement insurance unless you were planning to use it to buy genetic enhancement—as much enhancement as the insurance would pay for. To cover claims, therefore, insurers would have to charge as much in premiums as the cost of the enhancements, more in fact, since to stay in business, the insurance companies will need to cover their own overhead and make a profit. So there would be no point in buying genetic enhancement insurance; enhancement consumers would be better off just using the money they would spend on premiums to buy enhancement outright. The only advantage from having insurance might be the discounts that large insurance companies might be able to extract from the providers of enhancement products and services, like the lower fees that large managed care companies can negotiate with hospitals and physician groups. But enhancement consumers can attain similar savings by joining buying clubs or shopping at giant retailers. They can simply purchase their enhancements from Wal-Mart.

But even discount enhancement will be beyond the wherewithal of many segments of the population. For them, the final fallback would be the government. In order to ensure that everyone got access to genetic enhancement, the government could subsidize it. It could take its tax revenues and purchase enhancements for everybody. But the cost would be astronomical. The cost of universal access to in vitro fertilization alone, not counting the cost of accompanying genetic testing or manipulation, would be $120 billion a year, more than half the total federal budget for the Medicare program.[15] Giving everyone a modest subsidy to cover enhancements—say $10,000—would cost the government $2.5 trillion, almost twice the entire federal budget. The government might furnish a more modest subsidy, but this would provide access only to a fraction of the enhancements that wealthier individuals could purchase.

In short, it is inescapable that some people will be able to obtain genetic enhancement and some will not, at least not to a significant degree.

The societal implications of this certainty, as we shall see, are profound.

9

Inequality and Unfairness

Since genetic enhancements will not be available to everyone, those who can afford to purchase them with private funds will gain abilities that others can attain only with a great deal of time and hard work, or, more likely, not at all.

What happens when the enhanced, with their superior talents, interact with the unenhanced? This could occur in all sorts of situations: personal relationships; professional relationships like those between attorneys and clients or between patients and physicians; arm's length business relationships between employers and employees, landlords and tenants, lenders and debtors, sellers and purchasers. What happens when the enhanced and the unenhanced compete for the same scarce amenities or commodities—for employment, social status, political influence, affection, sexual favors, market power, money?

Business mogul Ted Turner is afflicted with the condition formerly known as manic-depression and now called bipolar disorder. He has controlled his condition with the drug lithium, which calms the frenetic highs during which his mind races at fantastic speed. But before the mammoth negotiations leading to the merger of Turner Broadcasting and Time Warner, it is said that he stopped taking the drug.[1]

Now imagine Ted Turner without the disabling effects of his condition. Before he goes into a business meeting, he swallows an enhancement pill or activates his internal drug synthesizer. The person on the other side of table, however, has no access to these products.

This is what would happen if genetic enhancements were available only to the well-to-do. The setting may not be on opposite sides of a conference table; anyone with access to Ted Turner would probably be wealthy enough to be enhanced themselves. But in stores and classrooms, in real estate offices and on used car lots, the enhanced would enjoy distinct and perhaps decisive advantages.

Many of these interactions will be zero-sum, in which one person wins and the other loses. In such competitions, the enhanced will be able to use their superior abilities and invariably they will be the ones who win. They will be positioned to take advantage of the less talented, to strip value from the unenhanced, to gain at their expense, to sell them shoddy goods, to cleverly betray their professional confidences. Even in interactions in which both parties come out ahead, the share of the enhanced person will always be greater.

In short, at an interpersonal level, genetic enhancement that is not available to all produces unfairness. The enhanced even might be called cheaters.

Wealth-based access to genetic enhancement doesn't only taint personal interactions, however. It threatens the very fabric of society. Since enhanced individuals will be able to replace determination and effort with a pill or with a genetic manipulation, they will be able to devote their time and energies to other pursuits, enabling them to pull away inexorably from everyone else. If, as is likely, enhancement confers abilities beyond what can be achieved with hard work alone, the enhanced will have opportunities beyond the grasp of even their most hard-working fellow citizens. Society will consist of the genetic haves and have-nots, and the haves will accrue greater social status, wealth, and power.

The social and political consequences could be grave. We could become a society of vast inequality. If genetic enhancement reaches anything like its full potential, the enhanced will be the most attractive, strongest, most graceful, most intelligent, most charismatic, and most inventive, and they will run the most successful businesses. All of these advantages will be rolled into the same persons. They will enjoy decisive advantages over everyone else in all realms of life—sports and beauty contests, game and talent shows, entertainment and the arts, admission to the best educational institutions, entry into the professions, political office and government appointment, getting rich or richer, and grabbing the most desirable mates. They will attain a monopoly over the best things in life, and their position at the pinnacle of society will be unassailable.

But is this dark image of the future realistic? Why would genetic enhancement be such a threat to society? People are not naturally equal now. Many are impoverished, lack a decent education, cannot obtain access to good health care, and do not participate in the political process. At the same time, some people are smarter than others, even much smarter. Some are born with talents that others lack and will never be able to obtain. Along with good luck and, occasionally, hard work, these talents give them enormous advantages. The richest one percent of Americans make more money than the 100 million poorest. We permit these people to enjoy the fruits of their natural talents and of their labors. We provide them with private schools, fancy houses, cars, and yachts, extravagant vacation resorts. We glamorize and drool over them on television.

Yet somehow our society manages to go on despite these frank inequalities, and our society, it is safe to say, is the envy of much of the world. Why would things be so different if genetic enhancement were only available to those who could afford them? Why would this constitute such a threat to our institutions?

It is possible that our society can handle the introduction of genetic enhancement. The technologies may turn out not to be very effective. The advantages they confer may not be substantial. But as we saw in Chapter 4, the development of enhancement technology is being borne on the crest of an unprecedented convergence of two revolutionary forces, biology and cybernetics. The prospects for developing truly powerful genetic enhancements, enhancements that give people unprecedented abilities, are too good for our concerns to be so easily dismissed.

Nor is it so clear that our society is weathering the present inequality of its citizens so successfully, even without the impact of genetic enhancement. The gap between rich and poor is widening steadily. Many of our inner cities are struggling to survive. Public school systems are turning out children who cannot read or write or do basic arithmetic. Voter apathy undermines and skews our elections. Campaign contributions from wealthy interests subvert the democratic process. Against this backdrop, genetic enhancement for the well-to-do could easily tip the scales toward instability and disorder.

One effect of genetic enhancement is that it transforms into acquired traits characteristics that formerly were derived from the natural lottery, over which individuals largely have no control. Instead of being born with a certain set of genes, people will be able to take drugs or gene infusions to produce desired abilities. Even children will no longer necessarily inherit a natural set of genes, but instead they will inherit genes that have been manipulated to give them enhanced capabilities.

At first, it may seem that this actually makes genetic enhancement less of a threat to society. Most people agree that no one deserves their natural talents. John Rawls states, "It seems to be one of the fixed points of our considered judgments that no one deserves his place in the distribution of native endowments, any more than one deserves one's initial starting places in society."[2] Ronald Dworkin argues that society has an obligation to rectify inequality that stems from what he calls "brute luck," which he defines as how risks fall out that are not "deliberate and calculated gambles."[3] But we generally allow people to enjoy the benefits of their acquired talents. If you strive to develop a skill, you get to keep the rewards. Therefore, if we consider genetic enhancements to be acquired, we might not mind so much the advantages that they conferred.

But this brings us back to the earlier discussion of authenticity in Chapter 7. There is an important distinction in our society between advantages acquired through hard work and practiced skill on the one hand, and through performance-enhancing drugs and genetic enhancements on the other. The former are earned and therefore admired and deserved. The latter are unmerited because of the manner in which they are obtained. They are acquired advantages rather than natural advantages, to be sure, but they are not acquired virtuously.

Given the expected cost of genetic enhancements, the most likely prerequisite for obtaining them will be wealth. But isn't wealth accumulated by hard work and practiced skill? Shouldn't people be entitled to the benefits from genetic enhancement that they purchase by dint of the sweat of their brows?

This may be true of enhancements purchased with wealth that has been acquired meritoriously, but that is not always the case. The wherewithal to purchase enhancements may be stolen, literally or figuratively, the former by robbing a bank, the latter by being a senior executive at Enron. Even such a champion of private property as Robert Nozick concedes that no one deserves possessions acquired by unjust means. Unjust means result, he says, when people "steal from others, or defraud them, or enslave them, seizing their product and preventing them from living as they choose, or forcibly excluding others from competing in exchanges."[4]

Even if the resources with which to purchase genetic enhancement were acquired by what Nozick would regard as "just means," neither they nor the advantages they conferred would necessarily be deserved. The resources may have been earned using talents that were conferred to a large degree by chance—by being born smart, or beautiful, or with perfect pitch—or through sheer luck. And in many cases, the wealth may not have been earned at all. It may have been inherited. So even if parents had worked hard and saved to give their children genetic enhancements, the children themselves would have done nothing to merit the resulting advantages. From a moral standpoint, their being enhanced is the same as being born with a natural assortment of "good" genes: It is merely good fortune, to be precise, the good fortune of having been born into a wealthy family. As philosopher Eric Rakowski points out:

So far as gifts from parents to children are concerned, the approximation to cases of unadulterated good brute luck generally seems quite close. Children do not choose their parents. And although rebellious offspring may forfeit their parents' love, to the extent that they enjoy it and profit materially from it they do so largely because, in a society based on the nuclear family, they were through no merit of their own better placed to win their parents' affection than were other people.[5]

In short, we face the prospect of people acquiring significant advantages through ill-gotten gains, or after having done nothing to deserve them. The former are surely forfeit, while the latter are no different morally than advantages acquired by accidents of birth. Even if we do not feel that people must lose their undeserved advantages because they do not deserve them, the more dramatic the advantages, the more unjust this state of affairs will seem to the rest of us, and the more resentment we will feel. At some point, this resentment could become so great that the unenhanced would become determined to deprive the enhanced of their privileges.

But won't we tolerate those people who are fortunate enough to become enhanced because their enhanced abilities will yield significant benefits for society as a whole? Won't they produce goods and services that will improve life for the unenhanced as well? Genetically enhanced inventors could design "better mousetraps" and enhanced business people could create better methods of distributing them to everyone. If it were our child who was stuck on the side of the mountain in the scene from the Introduction, wouldn't we welcome the arrival of an enhanced rescuer?

The notion that allowing people to enjoy the advantages conferred by undeserved talents is justified by the benefits to society is the rationale that meritocrats, libertarians, and free market theorists typically offer in support of social inequality. One of these theorists is John Gardner, a leading exponent of standardized testing for its ability to spot and reward native talent, as opposed to allocating slots at prestigious educational institutions on the basis of social pedigree or wealth. Gardner objects to what he calls "extreme egalitarianism" because it "signifies the end of that striving for excellence that has produced history's greatest achievements."[6] Even John Rawls would permit substantial social inequalities in order to increase total production of goods and services, so long as the least well-off were benefited along with the others:

> [T]he greater expectations allowed to entrepreneurs encourages them to do things which raise the long-term prospects of [the] laboring class. Their better prospects act as an incentive so that the economic process is more efficient, innovation proceeds at a faster pace, and so on. Eventually the resulting material benefits spread throughout the system and to the least advantaged.[7]

Good old trickle-down economics. Applied to genetic enhancement, the argument goes like this: Society would be better off allowing some people to purchase genetic enhancements for themselves and their children, even with the inequality that would result, because everyone would be better off, even though the enhanced will be even better off—indeed, astoundingly better off—than everyone else. The largesse created by enhancement will percolate downward, because the enhanced will share their advantages voluntarily with those less fortunate, or their gains will be redistributed.

Trickle-down is even offered as an explanation for why, in a democratic state, the majority does not rise up and seize more of the wealth of the privileged class.

As philosopher Thomas Nagel writes, substantial inequalities are "the price of efficiency."[8]

Of course, inequalities may become too extreme, even those that promote efficiency or that allow benefits to trickle down to those less or least advantaged. Inequality breeds envy. As John Rawls suggests, the resentment that inequality creates is not based so much on people's actual states, but on their perception of where they stand in comparison with those who are better off.[9] The resulting envy is both powerful and tell-tale; it is so valid an indicator of inequality that philosopher Ronald Dworkin offers an "envy test" to assess the success of his theory for achieving an equal distribution of resources: "No division of resources is an equal division if, once the division is complete, any [person] would prefer someone else's bundle of resources to his own bundle."[10] Envy is also highly unpleasant and corrosive. As Michael Walzer observes:

Opponents of the vision [of equality] often claim that the animating passions of egalitarian politics are envy and resentment, and it's true enough that such passions fester in every subordinate group. . . . But envy and resentment are uncomfortable passions; no one enjoys them; and I think it is accurate to say that egalitarianism is not so much their acting out as it is the conscious attempt to escape the condition that produces them.[11]

At some point, the disparity that the disadvantaged perceive between themselves and the privileged becomes so great, and their envy so strong, that the benefits to society that come trickling down seem merely like crumbs tossed from the table. At that point, the disadvantaged may no longer be willing to accept their inequality. They may take steps to rebalance the distribution of societal resources. This may not proceed smoothly. As one sociologist notes:

Inequality in the distribution of rewards is always a potential source of political and social instability. Because upper, relatively advantaged strata are generally fewer in number than disadvantaged lower strata, the former are faced with crucial problems of social control over the latter. One way of approaching this issue is to ask not why the disprivileged often rebel against the privileged but why they do not rebel more often.[12]

Moreover, it turns out that the promise of greater innovation and more goods and services for all is probably not the main reason why we tolerate so much frank inequality in our society. The main reason is that we believe not so much that everyone should have an equal share as that everyone should have an equal opportunity to acquire whatever share they can. In other words, what is important, what allows liberal democratic societies to accommodate the reality of capitalist inequality, is not equality per se but equality of opportunity. How many immigrants point with pride to how their children have succeeded in business or become professionals? How many poor parents work hard so their children will have the opportunities they can only dream of? Sociological surveys find that "[w]hereas most Americans are willing to tolerate sizable inequalities in the distribution of resources, they typically insist that individuals from all backgrounds should have an equal opportunity to secure these resources."[13] As John Schaar observes, the belief in equality of opportunity is instrumental in maintaining the prevailing social order:

> No policy formula is better designed to fortify the dominant institutions, values, and ends of the American social order than the formula of equality of opportunity, for it offers *everyone* a fair and equal chance to find a place within that order.[14]

Yet it is precisely the belief in equality of opportunity that is undermined by highly effective genetic enhancement. Those who obtain genetic enhancements will win the races and the talent contests, be awarded the Oscars, occupy the seats in Harvard's freshman class, become the doctors and lawyers, run the government, own the banks, and marry the girls and boys of their dreams. By virtue of genetic enhancement, they will do so invariably. The unenhanced will not be able to successfully compete. Nor will their children. They will not have any opportunity to succeed, except within the narrow limits reserved for them by the enhanced, and their opportunity will be equal only in terms of its absence.

This is what makes wealth-based access to genetic enhancement so pernicious. It destroys our belief in equality of opportunity, the scaffolding that props up our society of unequals.

Without this support, liberal democratic society will simply collapse. Perhaps not at first, or all at once. The genetic underclass might cede power for a time to its genetic superiors in return for the material benefits made possible by genetic advances. The members of the underclass might be content for a while with being upwardly mobile only within the confines of their class. The enhanced, in turn, might rule according to enlightened principles of *noblesse oblige*, taking care to permit sufficient benefits to trickle down to maintain political and social equilibrium. A democracy of sorts even might persist, with the unenhanced electing representatives who either were members of the genetic upper class or who were committed to preserving its privileges. Such a system might not look very different from our own in this respect, since we typically elect representatives who are considerably more privileged than their constituents.

However, this state of affairs, if it were to emerge, would likely be highly unstable. For one thing, the members of the genetic upper class would need to maintain a good deal of self-control to avoid overreaching. They would need to monitor and regulate each other to prevent anti-social excesses of greed. And a quasi-democratic system would be highly vulnerable to demagogues who achieved political power by promising to redistribute genetic enhancements more evenly.

The enhanced elite would be unlikely to accept being voted out of existence. It could be expected to employ all of its considerable talents to preserve its privileged status. The techniques it might use could include threats to withhold the fruits of genetic medicine from the unenhanced and overt interference with the democratic process. More subtly, the enhanced would be liable to control the media, which would allow them to control the outcome even of democratic elections in a manner far out of proportion to their numbers. Efforts by the unenhanced to maintain their political power might prove no more successful than current attempts to reform campaign finance laws in order to dilute the power of special interests.

In the end, we are likely to encounter an era of growing social chaos as society swings in ever-widening arcs between rule by underclass demagogues and rule by a genetic aristocracy. Eventually this could deteriorate into mob rule and finally anarchy. To rid itself of its underclass status, the unenhanced even might go so far as to try to destroy the scientific foundations of the genetic revolution, physically dismantling research centers and erasing mapping and sequencing data.

Alternatively, post-geno-revolutionary society could devolve into totalitarian rule by the genetically enhanced, who would employ whatever repressive techniques were needed to keep themselves in power and the unenhanced in check. As Aldous Huxley foresaw in *Brave New World*, these techniques might include the use of psycho-pharmaceuticals. To make the underclass more docile, the genetic elite might even use genetic manipulation.

This journey into nightmare will not happen overnight. If it occurred gradually enough, society would be given time in which to try to adapt. Unquestionably, then, the slower the pace of enhancement development, the better the chances for a successful adjustment. This is an important observation when it comes to designing a system to regulate genetic enhancement.

The typical scenarios in which society is capable of coping with genetic enhancement either assume that genetic enhancement will turn out to be a technological bust, which we dismissed earlier as overly optimistic, or they rely on the existence of a genetic middle class. According to this latter view, genetic enhancement, while expensive, will not be so costly that it is within the reach only of the very rich. A middle class will arise that can afford to purchase some forms of enhancement. The genetic aristocracy may even dole out enhancements to the middle class or give the middle class subsidies with which to purchase them. This population group forms a buffer between the unenhanced underclass and the genetic elite, enabling democratic institutions to survive.

The idea that the middle class facilitates democracy is as old as democracy itself. Even Aristotle was aware of this theory, asserting that "cities capable of being well governed are those sorts where the middle is large," and observing that "[d]emocracies are also more stable and longer lasting than oligarchies because of those in the middle, for the middle are more numerous in democracies than in oligarchies and have a greater share of honors."[15] Many contemporary political theorists credit the existence of a robust middle class for the ability of the United States and other Western liberal democracies to tolerate the huge inequalities between rich and poor. Kenneth Karst, in lamenting the growing disparities between rich and poor in the United States, comments, for example:

> The sense that the public order is legitimate—the belief on which a sound democracy must rest—depends on the sense that the individual has a stake in the system. That stake is both economic and social, and the social divisions I have recounted are very much influenced by anxieties over diminished economic opportunities. . . . [I]f large numbers of people should come to perceive themselves to be falling out of the middle class, we might not be laughing.[16]

But a middle class may not be capable of saving democracy from the loss of equal opportunity that attends wealth-based access to genetic enhancement. Enhancements, particularly the most effective ones, may still be too expensive for most members of the middle class to afford. The middle class itself could be divided into an upper middle class, which could afford enough enhancements to cause it to ally itself with the genetic elite, and a lower middle class, which would align itself with the genetic underclass. The crucial middle layer would disappear.

Even if a true genetic middle class persisted, it may not be able to prevent democracy from being overwhelmed by autocracy and totalitarianism. It certainly didn't prevent the rise of fascism prior to World War II. As Ronald Glassman states:

> [T]he commercial middle classes in and of themselves, even as a majority class, do not guarantee the rise of democracy or its maintenance. Structural conditions, such as the presence of a powerful feudal aristocracy and a functioning kingly bureaucratic state, can become important inhibitors to the democratic elective affinities carried by the middle class. Historically, circumstances of a crisis nature can turn the middle class away from democracy and toward some form of "nativist," conservative despotism, or towards a modernist form of despotism. Once such a despotism has been created, the middle class may not be powerful enough to overthrow it. It took WWII and the American victory, after all, to overthrow fascism and Nazism in Italy and Germany.[17]

Once again, then, although it is possible that the middle class will save society, this is by no means assured.

But we have not yet considered an even graver threat that genetic enhancement poses for Western liberal democracy. This is the prospect of germ line enhancement. Germ line enhancement, it will be recalled, alters the genetic endowment at an early enough stage of development that the changes are incorporated into a person's reproductive cells and passed on to their children. These children, starting out with the extraordinary genetic advantages conferred by their birth, can go on to amass even greater resources and privileges than their parents. These favors, along with any new advances in germ line enhancement techniques they install, will be inherited by *their* children. And so on.

Germ line enhancement creates a new aristocracy. A class whose privileges are not only unrivaled and unassailable, but inherited. An oligarchy which, having emerged from a society of ever-widening inequality, slams the door on the sustaining myth of equality of opportunity. In short, a genobility.

We have seen these kinds of societies before. World history is filled with examples. In medieval Europe, people were born within their social classes and there, with few exceptions, they remained. Upward mobility among peasants was restricted to infrequent cases in which religious education was obtained from a monastery, or a young man became apprenticed and eventually esquired to a knight. In slaveowning societies, persons who were born into bondage could be freed only at the pleasure of their masters or by escape, which risked recapture and severe punishment. Nor are these stilted, largely static societies just things of the past: The caste system in India is a constant threat to that nation's stability. Only recently, caste violence claimed thirty-four lives, mostly women, children, and old men, when gunmen identified as belonging to an "upper caste militia backed by rich landlords" stormed a lower-caste village in northeastern India.[18]

More than any other threat from genetic enhancement, it is the emergence of a genobility that is most incompatible with the Western liberal democratic tradition. A genobility would return us to a time when those privileged by birth owned virtually everything, set the standards of taste, and occupied political office. Where in place of one person—one vote, there was the right of the first night.

And as we contemplate the future of society in an era of genetic enhancement, let us not forget that modern democracy emerged only after the nobility was overthrown, typically in bloody revolutions.

10

Hubris

The prospect of powerful genetic enhancement technologies a-
vailable only to the well-to-do renders the future of society ex-
ceedingly bleak. Great rewards will go to those who can afford the
most extensive enhancements, rather than to those who earn by
determination and hard work. Under relentless pressure from employers and peers, anyone who can afford to become enhanced
will have no choice but to do so, and wealthy parents will invariably enhance their children to give them success in life. The genetically enhanced will use their exceptional skills to take advantage of the unenhanced. Equality of opportunity will be replaced
by the rule of the genobility. Democracy will die.

Even the technical problems of making sure that enhancement
is safe and effective may be resolved only at a significant social
cost. After all, who will serve as the guinea pigs? Not the rich and
famous, but the poor and vulnerable, who will be suckered into
participating in dangerous experiments in return for the elusive
prospect of enhancement advantages. Currently there are all sorts

of legal mechanisms designed to prevent this. Human experiments in most cases must be approved by government agencies and by review boards at the institutions that carry them out. Government regulations preclude offering subjects large sums of money in return for participating in clinical trials, or promising big breaks to prisoners. These restrictions were established in the wake of the abuses of the Nazis and of the Tuskegee experiments in the American South, in which uneducated, poor blacks suffering from syphilis were denied treatment in order to observe the progression of the disease. But the regulations could be repealed or ignored. One can imagine the argument: Enrolling poor people in genetic enhancement experiments will give them the only chance they'll ever have to obtain access to these amazing "wondergenes."

Still, we have not yet identified the gravest threat that genetic enhancement poses to the future of society. In the foregoing scenarios, human society persists, albeit in a grim and tyrannical state. People have experienced something resembling these conditions before, even if they provoke fear and despair.

But the ultimate expression of genetic enhancement could go far beyond anything we have known before. What if instead of just increasing height by several inches, genetic enhancement increased it by several feet? If someone could lift, not just a few hundred additional pounds, but a few thousand? If instead of being boosted by a score of IQ points, brainpower exceeded the capabilities of the fastest computers? As a result of the combination of cracking the genetic code and developing the technical means to modify it, technologies that will be perfected to prevent genetic disease and then extended to alter human traits give rise to the possibility that we will produce creatures with characteristics and abilities that transcend the bounds of *homo sapiens.*

In short, what if we produce creatures that are no longer human?

This and other potential consequences from genetic engineering motivate many persons to object to what they regard as "playing God." Jeremy Rifkin, head of the Foundation for Economic Trends and a long-time critic of genetic manipulation, warns that the "perspective [of scientists] is too narrow. . . .They don't think contextually; they're basic mechanics. We (the public) have to make sure that the technology is used intelligently. We attempt to play God, for good intentions. And then we get into trouble."[1]

Richard Land, executive director of the Christian Life Commission of the Southern Baptist Convention, similarly observes, "I think we're on the threshold of mind-bending debates about the nature of human life. We see altering life forms, creating new life forms, as a revolt against the sovereignty of God and an attempt to be God."[2] Michael Fox, a bioethicist from the Humane Society of the United States, states, "We are very clever little simians, aren't we? Manipulating the bases of life and thinking we're little gods." Genetic research, in his opinion, violates "the sanctity of life and may be regarded as an act of violence." "[T]he only acceptable application of genetic engineering," he says, "is to develop a genetically engineered form of birth control for our own species."[3]

But the results of genetic enhancement, it might be objected, could be extremely useful. They could go far beyond the rescuers in the Introduction. They could be marvelous people who could live and breathe underwater, who could survive in outer space, who could fly. If this is playing God, then so be it. After all, if you believe in God, who was it who gave us the ability to crack the genetic code, to develop the technology to manipulate the genome?

There is an old joke about a fundamentalist whose town was about to be engulfed in a flood. The police drove up and down the streets, urging everyone to evacuate, but the man refused to leave his home. "God will save me!" he assured the police. The waters began to rise, and a National Guard rowboat came by and yelled to the man to jump in. But he refused again. "God will save me!" he yelled back. Finally, as he was perched on the tip of his chimney, which was all that remained above the roiling waters, an Army helicopter hovered overhead and lowered a sling. Trusting in God, the man refused, and was drowned.

In heaven, and enraged, he marched up to Saint Peter to lodge his complaint. "I believed in God and here I am drowned! God didn't do anything!"

Saint Peter, looking surprised, replied, "Of course He did. Who do you think sent the police car, the rowboat, and the helicopter?"

In short, for those who believe that we are part of a divine scheme, who is to say that it isn't God who is giving us the tools to engineer our genetic futures? That He's passing the torch, so to speak, of creation?

For less theistically inclined teleologists, genetic enhancement could represent a transcendent evolutionary step in which mankind finally seizes its biological destiny from its former molecular masters, the genes. Hitherto, mankind, along with most other forms of life, has served primarily as a host for genetic parasites, who inhabit the cell nuclei and employ their hosts for reproduction. Having preprogrammed the degradation of its telomeres, they abandon the parent for its more useful offspring, leaving the former to die and be recycled. Genetic enhancement, then, appears as a mighty instrument of deliverance. It allows the host to seize control, to finally turn the tables on the genes and liberate itself from their tyranny.

Indeed, we have already begun to do this. We are producing forms of life that have never existed before. Bacteria that are engineered to consume and biodegrade oil spills. Chimeras called "geeps" that are half goat and half sheep.[4] A mouse that grows a human ear on its back.[5] Frogs and rabbits with glow-in-the-dark pigments. But the full fruition of genetic enhancement engineering will not be dumb, single-celled organisms or odd-looking little critters. It will be creatures that are articulate and self-aware, with minds that are highly—indeed fantastically—intelligent.

They will not be like us.

But since not everyone will be able to obtain genetic enhancement, some of "us" will still be around.

So what happens if we wake up someday and discover that the planet is inhabited by more than one sentient species?

This could be big trouble.

As the members of the superior species look down on the unenhanced, they may decide that the unenhanced, while clever little creatures, do not deserve the same civil and political rights. Driven by disregard or contempt for their unenhanced forebears, the superior species may become callous, malevolent slave-masters. They may even decide that they have no further use for the unenhanced. Except perhaps to toy with them.

In short, we could wake up someday and find that we have created a race of monsters.

The ancient Greeks had a term for this: for exceeding our grasp, for thinking that we could mimic the Olympian deities. They called it *hubris*. In *The Persians*, the playwright Aeschylus describes how the hubris of the Persian general Xerxes led him to construct a bridge of boats on which to march his army of invasion across the mighty Hellespont, only to have the waters rise up, destroy the bridge, and drown his soldiers. According to the Greeks, hubris, the delusion that one can achieve what is reserved for the gods, causes *hamartia,* or miscalculation. This in turn leads to *nemesis,* or catastrophe.

Obviously, the creation of a race of highly intelligent, genetically enhanced monsters would be folly of a grave sort. It would go far beyond the concern that genetic engineering deprives children of their "natural" genetic destinies, or that germ line manipulation would corrupt the gene pool so that humanity would be too homogenized to be able to confront a new environmental insult.

But it is not even the greatest folly. Nor is it entirely unfamiliar. After all, the Holocaust stemmed from the Nazis' belief that they were supermen and that Jews were subhuman.

No, the ultimate horror of genetic enhancement is not that we will create monsters. It is that we will create gods.

11

Solutions

The foregoing chapters describe a future in which genetic enhancement exacerbates the natural disparities between individuals, leading to the creation of a genetic aristocracy that invalidates the belief in equality of opportunity. Germ line enhancement cements the privileged status of the genobility. Democracy is a thing of the past. Society succumbs to tyranny or chaos as the genobility struggles to maintain its hegemony and the unenhanced strive to overthrow their rulers. The most frightening prospect is the creation of new sentient species of overlords if the changes wrought by genetic enhancement transcend the bounds of humanity.

What can be done to prevent this image of the future from becoming a reality?

One solution would be to distribute genetic enhancements to everyone, rather than only to those who can afford them. But when we considered this option in Chapter 9, we saw that the cost of providing genetic enhancement to everyone would be prohibitive. Do the math again. Even if enhancements cost only $10,000, it would cost over $2.5 trillion dollars to give them to everybody. This is almost twice the entire federal budget.

And that's just for the United States. The negative impact of genetic enhancement would be global. There could be a "gene drain" as countries vied to attract enhanced individuals, and the average citizens of poorer countries were left even farther behind in the race for material well-being. Even if genetic enhancement were somehow equally distributed within the United States and the other developed countries, peace and security would be threatened by the have-not regions of the globe. If people in those realms saw their chances for equality of opportunity irrevocably slipping away as the rest of the world pulled ahead genetically, they might feel that they had nothing to lose by attacking their more prosperous neighbors. The terror attacks of September 2001 would pale by comparison.

If a future with genetic enhancement is so bleak, then the obvious solution is to prevent people from becoming genetically enhanced, to ban it. A ban could target manufacturers of enhancement products, health care professionals who delivered enhancement services, and individuals who purchased, possessed, or used enhancements for themselves or their children. Congress and state legislators could make it a felony to sell, distribute, provide, or obtain enhancements.

This is similar to bans being proposed for human cloning. S. 1899, the Human Cloning Prohibition Act of 2002, sponsored by Senator Sam Brownback (R-Kansas) and others, would prohibit anyone from performing or attempting to perform human cloning, from participating in such an activity, from shipping and receiving a human clone in interstate commerce, or importing a human clone. The penalties are severe: imprisonment for up to ten years and a fine of not less than $1,000,000. Two bills sponsored by Democrats, S. 1758, sponsored by Dianne Feinstein (D-California), and S. 1893, sponsored by Tom Harkin (D-Iowa), ban the implantation and birth of a cloned human embryo but authorize human cloning for research and other purposes.

A variety of mechanisms could be employed to help enforce this ban. The FDA has asserted jurisdiction over enhancement products. For example, it licenses a manufacturer to market botulism injections for cosmetic use—to paralyze facial muscles in order to prevent wrinkles.[1] As noted in Chapter 5, the FDA already regulates liposuction machines, non-prescription contact lenses, and breast implants used for non-reconstructive purposes. Since somatic enhancements would fit the definition of a "drug" and genetic manipulations for enhancement purposes resembles gene therapy over which the agency exercises regulatory oversight, the FDA's assertion of authority to regulate genetic enhancements under the current provisions of the Federal Food, Drug, and Cosmetic Act is likely to be upheld by the courts, and if it is not, the agency can be given the additional authority by act of Congress.

As pointed out in Chapter 5, however, the FDA lacks the authority to regulate health care services, as opposed to products. This line is not always clearly drawn, as evidenced by the FDA's assertion of regulatory authority over gene therapy, which involves both a "product"—an infusion of genes, for example—and the "service" of introducing it into the body. Moreover, the FDA occasionally has conditioned the licensing of a product to restricting the health professionals who can furnish it to patients; for example, one product to treat urinary incontinence, known as the Contigen Bard Collagen Implant, can only be provided by "a physician specializing in incontinence who has had urology training in the procedure."[2] Still, in order to enable the agency to regulate genetic enhancement, it might be necessary for Congress to amend the food and drug laws to give the FDA clearer authority over the practice of medicine. This would encounter opposition from the American Medical Association, but their objections might be overcome by the perceived threat that enhancements pose to the fabric of society.

A bigger stretch would be to expect the FDA to enforce a ban on the possession or use of genetic enhancements. The agency has tended to focus its enforcement efforts on the purveyors of the articles under its control, rather than on the users or consumers of those articles. Its primary enforcement tool is to seize unlawful shipments, although it also has the authority to seek criminal punishment for violations of the laws within its purview, and to disqualify health care professionals, such as clinical researchers, from conducting experiments to be submitted to the agency in support of requests to approve new products.

To reach possessors and consumers of illegal enhancements, the more appropriate government bureaucracy would be the Drug Enforcement Agency (DEA). The DEA enforces the provisions of the Controlled Substances Act, which creates a system of "schedules" that categorizes products based on their perceived potential for abuse. This might be the most suitable regulatory framework for genetic enhancements, because while FDA regulation under the Federal Food, Drug, and Cosmetic Act is concerned primarily with products that pose a threat to their users, the Controlled Substances Act is the mechanism we designed to deal with drug products that pose a threat to society. Moreover, the DEA has plenty of expertise in regulating the manufacture of products under its jurisdiction and in interdicting illegal trade. The DEA regulates the entire process of production and distribution. It imposes ceilings on the manufacture of controlled substances, and even on the production of the basic chemicals that can be used in the manufacturing process. Genetic enhancements could be placed in Schedule I—which is reserved for products that have a high potential for abuse and no currently accepted medical use. For products in Schedule I, the DEA sets manufacturing limits at zero.

In fact, there already are precedents for a DEA-type approach to enhancements. In 1990, Congress placed anabolic steroids, which are abused by athletes for their muscle-building properties, in Schedule III of the Controlled Substances Act.[3] Schedule III is reserved for substances that have less potential for abuse than products on Schedules I and II, that have a currently accepted medical use, and the abuse of which may lead to moderate or low physical or high psychological dependence.[4] While the restrictions the DEA imposes on Schedule III substances are not as strict as the complete prohibition on substances in Schedule I, the placing of steroids on any schedule shows that the government is thinking along the lines of strict regulation of enhancement products.

An even clearer analogy stems from congressional action in connection with human growth hormone (HGH). In 1991, Congress amended the Federal Food, Drug, and Cosmetic Act to make it a federal felony to distribute or possess human growth hormone—

which is made with recombinant DNA technology—"for any use in humans other than the treatment of disease or other recognized medical condition."[5] Here, the law specifically targets enhancement uses of a therapeutic genetic substance. In effect, this law is the first federal ban on genetic enhancement.

A threshold question is whether Congress has the power to enact such a law. In other words, is it constitutional? What would the Supreme Court say if the ban on using human growth hormone for enhancement purposes were challenged in court as an impermissible infringement on personal freedom? Although there have been no such challenges yet, they can be expected in the future from a variety of sources if the government adopts an across-the-board prohibition on genetic enhancement. Challenges would come from individuals who wish to enhance themselves, from parents who seek to enhance their children, from health care professionals and provider institutions who hope to make money selling enhancement services, and from the manufacturers and distributors of enhancement products and equipment.

The success of these lawsuits will depend on the nature of the enhancement technologies that the plaintiffs sought to sell or employ. The Supreme Court has recognized the government's constitutional authority to constrain individual choice in connection with the use of therapeutic drugs. For example, the Court upheld the FDA's seizure of laetrile, an unapproved drug made from apricot pits that was touted as an anti-cancer agent, in the face of complaints by patients and physicians that the FDA was depriving them of the freedom to decide which treatments to employ. The Court even brushed aside the argument that patients ought to be allowed to take the drug when all other treatments had failed.[6] It is unlikely that the Court would interfere with a government effort to restrict the use of a drug for non-therapeutic purposes, particularly if the government argued that the drug posed a safety risk to its users or a threat to society.

A more difficult constitutional question would arise in the case of genetic enhancement of children by their parents. The courts have recognized that parents have a broad, constitutionally protected right to raise their children as they wish. In *Cleveland Board of Education* v. *LaFleur,* the Supreme Court stated that "freedom of personal choice in matters of family life is one of the liberties protected by [the Constitution]."[7]

In *Moore v. City of East Cleveland*, the Court observed that "[t]he Constitution protects the sanctity of the family precisely because the institution of family is deeply rooted in this Nation's history and tradition."[8] These and other cases led a federal court of appeals judge to say that "the Constitution mandates a judicial obligation to respect, if not protect, parental authority in decisions relating to general family matters. . . ."[9] Specifically addressing the rights of parents to make decisions for their children, the Supreme Court has said that "the values of parental direction of the . . . upbringing and education of their children . . . have a high place in our society," that parents have a right, protected by the due process clause of the Constitution, "to direct the upbringing and education of [their] children . . . [and that this right could] not be abridged by legislation which has no reasonable relation to some purpose within the competency of the State."[10]

At the same time, however, the courts also have firmly acknowledged that the state has the right to interfere with parental decisions that may cause harm to their children. In a case called *Parham v. J.R.,* involving whether parents could have their child voluntarily committed, the Supreme Court noted that "a state is not without constitutional control over parental discretion in dealing with children when their physical or mental health is jeopardized."[11] For example, the courts have routinely refused to hold unconstitutional laws that protect children from parental abuse and neglect,[12] even laws that are quite broad and were challenged on the ground that they were impermissibly vague.[13] The state can even interfere when parents use corporal punishment on their children. In Florida, for example, parents cannot be punished under the state's felony child abuse law for corporal punishment to discipline their children, but the parents can still be punished for simple child abuse when a beating sends a child to a hospital.[14]

If the enhancements were physically dangerous, therefore, the courts would likely uphold a ban on enhancing children as a permissible exercise of the government's police power to prevent harm to the children. Parents might contend that the benefits of enhancement outweighed its risks, and the courts would have to balance the risks and benefits against the traditional rights of parents described above. The approach the courts would be most likely to take is something like the following, which Judge Aldisert adopted in a case concerning whether parents could voluntarily commit their child to a state institution:

> [P]arents have a substantial constitutional right, as head of the family unit, to direct and control the upbringing and development of their minor children. If the parental decisions amount to abuse or neglect of the minor child then the parental right is no longer constitutionally protected, and the state, as parens patriae, may intervene to protect the child. Absent a showing of abuse or neglect, however, the parental right remains substantial and may be subject to governmental interference only when such interference is supported by a significant governmental interest.[15]

Under this test, the answer to the question of whether parents have a constitutional right to enhance their children genetically will depend on whether the courts find that genetic enhancements cause net harm to children, and if so, if the courts conclude that the harm is substantial enough to be regarded as abuse and neglect.

A ban on the use of genetic enhancements as part of the reproductive process would be especially hard to sustain. The Supreme Court has recognized a constitutional right to procreate, and it has ruled that a government restriction can be upheld only if it fulfills a compelling state interest. In one case, *Skinner* v. *Oklahoma,* the Court stated that a statute that provided for compulsory sterilization of criminals "involves one of the basic civil rights of man.[16] Marriage and procreation are fundamental to the very existence and survival of the race." More recently, the Court reiterated its recognition of "the right of the individual, married or single, to be free from unwarranted government intrusion into matters so fundamentally affecting a person as the decision whether to bear or beget a child."[17] In cases involving contraceptives and abortion, procreative liberty therefore has trumped state efforts to restrict the right not to have a child.

Still, the government could argue that a ban on genetic enhancement did not restrict the right to have a child, but only the right to have an enhanced child. The issue then would turn on the previously discussed question of whether the government legitimately can prevent parents from enhancing their children.

Even if a ban on enhancing one's children interferes with the fundamental right to procreate, the ban still can be upheld as constitutional if it fulfills a "compelling state interest." The Supreme Court has explained that a "compelling state interest" is limited to "only those interests of the highest order and those not otherwise served,"[18] but has failed to set forth a general test for determining when this is the case. Like pornography, in Justice Stewart's famous words, the courts know it when they see it.[19] Thus, they have "seen it" in the need to prevent and catch burglars, to prevent the publication of certain representations of currency to prevent counterfeiting, in the need to restrict campaign contributions, and so on.[20] But if anything is a compelling state interest, avoiding or reducing the threats posed by genetic enhancement, and especially by germ line enhancement, ought to count. What could be more compelling than preserving democratic institutions or the fate of the human species itself?

In terms of upholding a ban on genetic enhancement, the question would be how clear its dangers could be made to appear. The courts might decline to support a ban based on speculation about remote, future harms. But by the time the danger became clear and present, a government ban might be too late.

One way a government ban might avoid the harshest forms of constitutional scrutiny would be to target non-traditional methods of reproduction, like in vitro fertilization (IVF). IVF most likely would be a necessary precursor to the most effective enhancements—enhancements that were accomplished by actually manipulating the genes—as well as to germ line enhancement. Congress could forbid health care professionals and providers from furnishing enhancements as part of the IVF process. Since they are not natural forms of procreation, IVF and other forms of assisted reproduction might not seem to be entitled to the same broad protection under the Constitution.

However, the few court cases that have dealt with government restrictions on non-coital forms of reproduction like IVF appear to recognize that, like traditional procreation, they too deserve broad constitutional protection from state interference. A federal court in Illinois, for example, struck down that state's prohibition on fetal research on the ground that, by interfering with choices made during assisted reproduction, it interfered with a woman's right to privacy. "It takes no great leap of logic," the court stated, "to see that within the cluster of constitutionally protected choices that includes the right to have access to contraceptives, there must be included within that cluster the right to submit to a medical procedure that may bring about, rather than prevent, pregnancy."[21] And in *Cameron* v. *Board of Education,* an Ohio federal court stated simply that "a woman has a constitutional privacy right to control her reproductive functions. Consequently, a woman possesses the right to become pregnant by artificial insemination."[22]

It is even less likely that the Supreme Court would uphold a congressional ban on passive forms of reproductive genetic enhancement—for example, decisions by couples about which embryos to implant as part of the IVF process. So long as IVF itself remained legal, a law that prohibited parents from selecting which fertilized embryos to implant based on the results of genetic tests for non-disease traits would put the government in the position of attempting to control the kinds of children parents could conceive. The government would be interfering directly with the parents' procreative freedom. Insofar as it restricted the parents' ability to decline to bring a fertilized ovum to term, such a law would resemble a prohibition on abortion at the earliest possible stage of pregnancy—something that the Supreme Court has made clear is beyond the government's constitutional power.

Only if the selection for enhancement purposes somehow significantly harmed the resulting fetus might government regulation be permissible. This would be hard to show, since all the parents would be doing is choosing the most positive among naturally occurring embryonic traits. Moreover, it would be difficult to embargo passive enhancement while at the same time permitting the IVF process itself to take place, since the most obvious harm is to the embryos that are discarded, a harm which is inherent in the IVF process.

Of course, there is an obvious solution if government attempts to ban genetic enhancement were deemed unconstitutional: amend the Constitution. This is a difficult process, typically requiring a bill to pass both houses of Congress by a two-thirds majority, followed by ratification by three-fourths of the states within a certain number of years. But if genetic enhancement were viewed as a serious enough threat to society, the political will might be found to make the necessary changes. The problem would be to avoid enmeshing an enhancement amendment in efforts to ratify a pro-life amendment, an amendment that restricted parental decisions outside of the enhancement context, which would be unlikely to be adopted. To avoid this, the constitutional provision would have to be crafted so that Congress could ban genetic enhancement while parents' procreative liberty in other contexts was preserved. For example, the amendment would have to distinguish between enhancement and therapeutic choices, so that parents would still be able to decide not to bear children with impairments, and the amendment would have to recognize a woman's constitutionally protected right to abortion.

Even if constitutional impediments were removed, a ban on genetic enhancement would face enormous practical difficulties. Shipments and sales of enhancement drugs would have to be detected and interdicted. In order to control the enhancement of children during the process of reproduction, the government would have to involve itself in hitherto private decisions between parents and health care professionals.

Undoubtedly the government and private religious, professional, and public interest organizations would mount a major public education program to try to convince individuals to eschew genetic enhancement for themselves or their offspring. The program might play upon the reality, or upon fears, that enhancement was neither safe nor effective. As part of a ban, the government would refuse to permit the marketing of products for enhancement uses, which means that there might be little scientifically reliable information about how safe or effective they were for those uses. (Safety and efficacy studies might be conducted in other countries that did not ban the technologies, but the government's education campaign might seek to discount those studies if they were not carried out according to U.S. scientific standards.) The government could point to the dangers of genetic enhancement to society as well as to the individuals themselves. One could imagine a vigorous "Just

Say No" campaign, featuring as spokespersons leading citizens who rejected genetic enhancement for religious, ethical, or political reasons, and peppered with horror stories about victims of failed enhancement attempts.

Another approach would be to rely on health care professionals to police a ban on genetic enhancement themselves. Many forms of genetic enhancement would require the assistance of health care professionals: prescribing approved drugs for off-label enhancement uses, supplying genetic testing, and providing IVF services for passive reproduction and active preimplantation genetic manipulation. These health care professionals may be members of organizations that issue ethical pronouncements to govern professional behavior. The American Medical Association, for example, has a Council on Ethical and Judicial Affairs that produces a Code of Ethics. Similar canons may be issued by the American College of Physicians, by specialty societies like the American College of Obstetrics and Gynecology, and by organizations of genetics professionals like the American Society of Human Genetics, the American College of Medical Genetics, and the National Society of Genetic Counselors. Violating these principles can lead to censure and expulsion from the group, which sometimes can be used against the professional in a court of law, for example in a malpractice case.[23] Professionals may be influenced by the ethical views of respected colleagues in influential medical journals; recall that an apparently effective voluntary moratorium on recombinant DNA research followed a call for one by Paul Berg and others in the journal *Science.*

One professional organization, the American Medical Association, already has taken a position on genetic enhancement. In 1994, the association's Council on Ethical and Judicial Affairs published its views as part of a report on prenatal genetic screening. The council observed that efforts to enhance "desirable" characteristics through the insertion of a modified or additional gene, or efforts to "improve" complex human traits, ran contrary to the ethical traditions of medicine as well as the egalitarian values of society.[24] The report went on to state that "genetic interventions to enhance traits should be considered permissible only in severely restricted situations," namely, circumstances that fulfilled all of the following three conditions:

(1) "clear and meaningful benefits to the fetus or child";
(2) "no trade-off with other characteristics or traits"; and
(3) "equal access . . . irrespective of income or other socio-economic characteristics."

The second condition is perplexing. Why should parents be prohibited from deciding, for instance, that a significant increase in their child's strength was worth a slight decrease in dexterity? But the statement nonetheless represents an example of how professional self-regulation might play a role in limiting access to genetic enhancements.

How likely are physicians and others to obey professional calls to refuse to provide genetic enhancements to those who seek their services? Some may not agree with the ethical views of professional organizations. Not every member of the AMA, for example, is likely to endorse the position that genetic enhancement should be provided only if it is available to everyone. Moreover, physician incomes are being squeezed by managed care and by rising malpractice premiums. Genetic enhancement is certain to be highly lucrative. Since health insurers won't be covering it under their policies, enhancement providers will be able to avoid the hassles of dealing with the claims paperwork and the second-guessing of managed care plans. Like cosmetic medicine, enhancement services will be furnished on a cash-on-delivery basis. Under the circumstances, health care professionals may be unable to resist the economic lure of the enhancement business.

In short, if society resolves to impose an effective ban on genetic enhancements, it will have to supplement individual and professional self-control with governmental regulatory measures. Congress could make it illegal to possess, use, or provide genetic enhancements. The DEA, together with state and local police authorities, could be charged with searching out and punishing violators, including physicians and clinics that furnished enhancement services as part of IVF and other types of assisted reproduction. Health care professionals who violated the laws could be fined or imprisoned, and they could have their professional licenses suspended or revoked. Hospitals and other facilities that performed enhancement procedures could lose their operating licenses, their accreditation, and their privilege to receive Medicare and Medicaid reimbursements from the federal government.

Other restrictions would target genetic testing for enhancement purposes. This would require a major overhaul of the current regulatory framework. At present, responsibility for regulating genetic testing is divided among three federal agencies—the FDA, the Centers for Medicare and Medicaid Services (CMS), and the Centers for Disease Control and Prevention (CDC). The FDA approves tests for accuracy and reliability, but only those sold as "kits" to laboratories. (The FDA also would have authority over genetic tests that could be performed in the home, like home pregnancy tests. At present, no such genetic tests are available.) CMS inspects and certifies laboratories. The CDC oversees testing services provided by non-commercial labs. A big loophole is genetic testing services performed by commercial laboratories using their own testing process, often based on a genetic sequence that the lab has isolated, purified, and patented. While the lab itself must be certified by the CMS (at least if it provides testing services to Medicare or Medicaid patients), no government agency has responsibility for evaluating the accuracy and reliability of the tests themselves. Another loophole applies to laboratories like those in some hospitals which are exempt from federal oversight because they perform such a small number of tests in any given year. To effectively control access to genetic testing for enhancement purposes, these loopholes would have to be closed and a single government agency given the responsibility for regulating the genetic testing industry.

The FDA would have a crucial role to play in the enforcement of a ban on genetic enhancement. The agency would refuse to license products for genetic enhancement purposes. Since the FDA does approve some products for cosmetic uses, like breast implants, non-prescription contact lenses, liposuction machines, and botulism injections to temporarily remove wrinkles, it would need to distinguish between permissible enhancements and banned genetic enhancements. Assuming that the FDA will continue to approve non-genetic cosmetic interventions, a ban on genetic enhancement would require the FDA to be able to distinguish between genetic and non-genetic products. This may not be so easy. Agency officials might be able to identify products that employed genetic engineering, such as drugs made with recombinant DNA techniques or that consisted of infusions of genetically altered DNA, but what about enhancement products produced with traditional manufacturing techniques that were made possible by insights from genetic science, like proteins or protein-blockers developed by sequencing the stretches of the genome that code for non-disease characteristics? How much genetic information would researchers have to rely upon in order for the FDA to place a product in the banned category, and how would agency regulators determine that genetic knowledge had in fact been employed in product discovery and development? The alternative would be for the FDA to refuse to approve only those enhancement products made with genetic engineering outright, but that might allow many other enhancement products to be lawfully marketed, undermining the effectiveness of the ban.

Another problem for the FDA mentioned earlier is its historic lack of jurisdiction over the practice of medicine, as opposed to the products used by medical practitioners. This would complicate FDA efforts to ban enhancement practices, like gene insertion and deletion for enhancement purposes or, assuming it was permitted, passive reproductive enhancement, involving the identification and implantation into the womb of embryos selected for their desirability, or the aborting of undesirable fetuses following genetic testing for enhancement purposes. To overcome this deficiency in the FDA's authority, Congress would have to change the laws to give the agency the power to regulate the practice of medicine.

A more serious impediment for the FDA is physicians' current ability to lawfully prescribe drugs or to use devices for purposes for which they have not been approved. This so-called "off-label" or "unapproved use" practice is widespread. For example, until recently the FDA had only approved Botox, the botulism injections reported to be widely used to prevent wrinkles, for treating spasmic disorders of the eye muscles.[25] In a sense, most drugs prescribed for women or children are provided in a manner not approved by the FDA, since virtually all such drugs have been tested only in adult male subjects, and the agency has reviewed no safety or efficacy data supporting their use in female or pediatric patient populations.

Another potentially significant loophole in the FDA's ability to regulate genetic enhancements stems from its limited authority over so-called dietary supplements. Under the Dietary Supplement Health and Education Act (DSHEA), the FDA can require manufacturers of these over-the-counter products to place disclaimers on the labels stating that the product is not approved by the FDA, and the agency has some ability to control products that make overt claims that they treat or cure disease.[26] Otherwise, however, manufacturers are free to market these products without obtaining a government license, and they could use this freedom to sell enhancement drugs.

For the FDA to be able to halt enhancement uses of products approved for therapeutic purposes, Congress would have to expand the FDA's authority to enable it to prohibit off-label enhancement use by physicians, and to more closely regulate dietary supplement products. This would embroil the agency in regulating products because of concerns about their social impact, rather than merely because of traditional safety and efficacy considerations. The FDA would need both additional congressional authority and additional expertise in social and ethical fields.

Even with expanded authority and expertise, however, the FDA would play a modest role in enforcing a government ban on genetic enhancement. The bulk of the responsibility would fall on a police-type agency like the DEA, backed up by local law enforcement. If the dangers posed by genetic enhancement were regarded as severe, as well they should be, we could expect a "war on genes" similar to the war on drugs.

But a war on genes would not only mimic the techniques the government employs in the war on drugs. It is bound to echo the failures. The more powerful and attractive the genetic enhancement, the more unable the law will be to achieve a total ban. In the face of attempts to make genetic enhancement illegal, a black market is bound to emerge.

This is confirmed not only by the widespread inability of the government to prevent people from using illegal recreational drugs, but by past experience with the closer analogy of performance-enhancing drugs in sports. As described earlier in Chapter 7, this has been a continuing headache for athletic organizations. In 1997, a reporter for the *San Diego Union-Tribune* wrote that these drugs "were seeping through the sports world like an injectable steroid is absorbed into the blood stream."[27] The only people being caught, observed the reporter, were either "poor or stupid." During the most recent Olympics, competitors who had failed drug tests were being stripped of their medals even on the last day of the games. Despite tremendous international and domestic efforts, athletes continue to be able to obtain supplies of illegal enhancement substances.

Of course, not all forms of genetic enhancement would involve merely taking a handful of pills or a series of injections. The most powerful enhancements might require more exotic techniques that relied on more sophisticated medical equipment and skilled providers. But here, too, black markets could flourish, as illustrated by the U.S. experience with abortion. Before *Roe* v. *Wade* legalized abortion in the United States by declaring a complete ban unconstitutional, not only could you go to Mexico, where abortion doctors were eager to accept Yankee dollars, but it was possible to obtain "back alley" abortions domestically if one "knew the right person." Indeed, it is estimated that, in the early 1960s, one out of every five U.S. pregnancies was terminated by an abortion, even though virtually all the procedures performed in the United States were illegal. In 1962 alone, over one million abortions were believed to have been performed, half of them by physicians.[28] Obviously there were quality problems: a number of women died as a result of botched techniques. But the wide availability of abortions demonstrates the difficulty of enforcing a complete ban even on technically demanding procedures such as genetic testing and passive reproductive enhancement using IVF.

Apart from the willingness of unscrupulous professionals to violate a ban for the right price, even honest physicians might have difficulty distinguishing between legal and illegal uses of enhancement products. Consider the statutory prohibition on enhancement uses of human growth hormone. As noted earlier, distribution other than for the "treatment of disease or other recognized medical condition" is a federal felony under the Federal Food, Drug, and Cosmetic Act. But when it comes to stature, what is a disease or medical condition? The terms are not defined in the law. Extreme forms of enhancement use could perhaps be clearly identified, like when parents seek the drug for already tall children in the hopes of making them future NBA material. But what about children who, while still "normal" (which, it will be recalled from Chapter 4, is defined arbitrarily as within two standard deviations of the population mean), are at the short end of the spectrum? What about children in the middle of the normal range whose parents believe that they would have a better chance at success if they were a little taller?

These problems are familiar aspects of the war on drugs. In enforcing the Controlled Substances Act, the DEA has to try to distinguish between legal and illegal uses of controlled substances that, like anabolic steroids, are placed on Schedules II through V because they have some legitimate medical uses. This is clearly not easy to do. State and federal drug enforcement officials have had only limited success in stemming the flow of prescription drugs onto the illicit market; approximately 27 percent of illicit drug use in the U.S. involves prescription drugs.[29]

If the threats from enhancements were perceived as serious enough, and enforcing a ban were made sufficiently difficult by the problem of controlling off-label enhancement uses of products that were approved for therapeutic purposes, the government could refuse to approve products altogether. Not only enhancement uses but distribution for any purpose, therapeutic included, could be made illegal. This would be an extraordinary step. It is virtually inconceivable that the FDA would refuse to license a drug to treat a widespread and devastating disease like Alzheimer's, for example, because of the risk that the drug might be used as a cognitive enhancer by healthy individuals. Only products that presented the gravest risks from enhancement use, and that offered relatively modest therapeutic value, might be regulated in this fashion.

Another bold measure would be to block research on enhancement products and procedures. One approach would be to prohibit government funding of enhancement research. Notwithstanding

Celera's privately funded accomplishments, there is no question that NIH funding greatly accelerated the mapping and sequencing of the human genome. Denying federal funding for enhancement research would significantly slow the pace of enhancement development. There is a good deal of precedent for this as well. For years, the Recombinant DNA Advisory Committee at the NIH refused to entertain proposals for somatic cell human gene therapy experiments,[30] and the government still will not fund research that proposes to alter the germ line. Similar government restrictions have prohibited NIH funding for research using fetal tissue or embryos.[31]

But government funding restrictions only affect government-sponsored research. As the Celera story demonstrates, private companies are also capable of achieving significant research breakthroughs, and the potential demand for genetic enhancements would give them ample financial incentives to make substantial investments in enhancement research. A research ban would have to extend to these activities as well. If the FDA refused to license enhancement products, manufacturers would have less incentive to privately finance expensive efficacy and safety research. Congress could go even further and expand federal funding prohibitions to make it illegal to conduct private research on enhancements. The government could enlist the aid of the scientific community, calling for a voluntary moratorium on enhancement research like the moratorium on recombinant DNA research in the early 1970s described in Chapter 5.

One important tool in banning enhancement research would be to deprive the resulting discoveries of patent protection. Congress could amend the patent laws to prohibit patents from being issued for enhancement inventions. Unpatentable discoveries would include the identification, isolation, and purification of the genetic sequences that code for non-disease characteristics; enhancement drugs produced through genetic engineering, including recombinant DNA technology; and otherwise patentable components of genetic tests for enhancement purposes, like laboratory reagents and equipment. The result would be to deny companies and their investors the ability to exclude others from appropriating their inventions and selling them as their own, or using them to develop other patentable downstream enhancement technologies. Lacking patent protection, companies would be less willing to invest substantial sums of money in enhancement research.

Patent protection, however, is not the only way inventors can attempt to exclude others from their discoveries. Instead of filing for a patent, which requires inventors to reveal their discoveries to the world so that others can use their knowledge to make further progress in the field, researchers can keep their discoveries private, relying on old-fashioned trade secret status to protect them from the prying eyes of would-be competitors. Many lucrative inventions are protected in this fashion, such as the formula for Coca-Cola. Trade secret status is not an absolute guarantee of exclusivity, however; competitors can come up with the invention independently, and they can "reverse-engineer" the end product to figure out how it was made and what its components consisted of.

One problem with all these suggestions is the potential for off-label use. As noted earlier, private companies could avoid researching the enhancement uses of their products and still count on sales from off-label uses of products that the FDA had approved for therapeutic purposes. Even if the government prohibited off-label use, the products and services might be readily available on the black market.

But there is an even bigger problem with all these attempts to control genetic enhancements. So far, all we have been discussing are restrictions imposed by the U.S. government. What is to prevent individuals seeking genetic enhancements from obtaining them outside the country? When abortions were illegal in the U.S., pregnant women traveled to Mexico. Why won't people simply travel to Mexico to purchase genetic enhancements? This practice even has been given a name before it has become a reality: "genetic tourism."[32]

Foreign access to genetic enhancements could take a number of forms, depending on the enhancement in question. In the first place, individuals could attempt to import into the U.S. illegal enhancement drugs that were manufactured abroad. A similar practice arose in the past with AIDS drugs that were available in other countries but were still going through experimental stages in the U.S.; patients even formed underground "buyers clubs" to secure supplies of these foreign products.[33] Obtaining enhancement products from foreign sources would be facilitated by the Internet. A casual search of the World Wide Web reveals numerous sites, the vast majority presumably overseas, that offer to ship drugs anywhere in the world, sites like the "International On-Line Pharmacy Club" that promises to teach you "the secrets of buying prescription drugs without a prescription on a 70% discount"; "Diet Doc," which adds a $75 on-line physician consultation charge to its drug shipment costs; and "WorldPharmacies.com," which advertises drugs not available in the U.S., such as "sports enhancements," for $4 and up. One writer reported that within two months of trying to obtain drugs over the Internet, she collected "30 capsules of Xenical, 30 tablets of Prozac, 100 Ultrams, 100 penicillin, and Preven—a kit with four birth control pills to prevent pregnancy after a sex slip-up [and] five bottles of injectable Xylocain."[34]

Instead of having enhancement drugs delivered to them from abroad, individuals could travel abroad themselves, buy the drugs, and return with them to the United States. U.S. Customs regulations permit people to bring up to fifty dosages of controlled substances back with them into the United States, even without a valid U.S. prescription, so long as the drugs are declared, for personal use, and in their original container. However, the rule only applies to drugs that are approved by the FDA for sale in the United States; accordingly, genetic enhancements would be contraband.[35]

Interdicting the importation of illegal drugs into the U.S. is a central goal of the war on drugs, and presumably the same techniques which the DEA, U.S. Customs agents, and postal officials use in their attempt to seal the border would be employed in a war on genes. But some somatic enhancements may be long-lasting enough that they would continue to produce enhancement effects for some time after they were taken. It would be much more difficult to detect individuals who traveled abroad, ingested or infused enhancement products, and then returned. Athletes might go abroad, use an enhancement drug to build up muscle mass, and then stop taking the drug; they then could return to compete in the United States without risking being caught with illegal drugs in their possession. There have been many instances of individuals who have obtained drugs in other countries that were illegal in the

United States. A classic example is laetrile, a substance made from apricot pits that was supposed to treat cancer. When the FDA seized U.S. supplies of the unapproved drug, patients, including the actor Steve McQueen, journeyed to Mexico.[36] (McQueen died shortly thereafter.) The same thing can be expected to occur with somatic genetic enhancements. For example, U.S. citizens could travel abroad and have genetically modified cells inserted into their bodies. Like the cells infused into Ashanti DeSilva, these may continue to produce enhancement proteins for some time before being sloughed off by the body. Alternatively, the art of somatic genetic engineering may progress to the point that infused cells are capable of reproducing and creating their enhancement effects indefinitely. Ultimately, it may be possible to go abroad and manipulate the DNA of embryos and early-stage fetuses, yielding germ line enhancement modifications. Women pregnant with genetically engineered fetuses could return and give birth in the United States, or they could bear the children overseas and then return with their enhanced offspring.

An important contrast between a war on genes and the war on drugs would be the attitude of foreign governments. Foreign governments either support the war on drugs or claim officially that they do. While corrupt officials may hamper efforts to combat illegal drugs, the official stance of their governments obligates the organs of state to cooperate with the United States, or at least to appear to cooperate.

Many countries undoubtedly would join the United States in adopting and enforcing a ban on genetic enhancements. European nations are already well ahead of the United States in imposing restrictions on certain forms of biotechnology, such as germ line therapy; the European Parliament passed a resolution in 1996 banning human germ line modifications.[37] Other advanced industrial nations may follow suit. But poorer, developing countries may view the development of a genetic enhancement industry as a potential gold mine, capable of attracting foreign investment in biotechnology and foreign exchange as individuals flock across their borders to secure enhancements that are forbidden in their home countries. These nations may refuse to ban enhancements domestically or to cooperate with the efforts of the United States and others to restrict access to enhancements abroad. Instead, developing countries may welcome and perhaps even subsidize the establishment of enhancement production and provider facilities. They could even view the genetic enhancement of their own citizens as a solution to national impoverishment, since this would give these individuals advantages that would enable them to compete more effectively against their counterparts in more advanced countries. Of course, if genetic enhancement is too expensive to be widely available in wealthier countries, it is not likely to be readily accessible to the average citizen in poorer nations, but the wealthy elites of those nations may seize upon genetic enhancement as a rapid method of increasing their wealth and power. This will draw them farther apart from their compatriots, but also from elites in countries that ban genetic enhancement.

But perfecting genetic enhancements will require a certain degree of technical sophistication. Where will developing countries obtain the scientific and medical expertise necessary to create a genetic enhancement industry? They may want to send their citizens to study at research institutions and biotech companies in other countries, but these countries would be bound to establish controls to prevent foreigners from studying enhancement techniques on their soil and then returning home. Acquiring the necessary expertise wouldn't be that difficult, however. After all, if industrially advanced nations forbid enhancements, scientists eager to research and develop genetic enhancement technologies would emigrate to countries willing to allow them to pursue their private research agendas. As mentioned in Chapter 5, Martin Cline attempted to perform gene therapy on patients in Italy and Israel while waiting to obtain approval to conduct his experiments in the United States. In January 2001, an American infertility specialist, Panayiotis Zavos, announced that he and an Italian partner, Severino Antinori, who had become famous for helping a sixty-two-year-old woman to give birth by IVF, were forming a company at an undisclosed Mediterranean location that would clone human beings.[38] A Canadian cloning group, the Raelians, similarly declared their intent to begin cloning humans overseas.[39]

The establishment of "safe havens" for genetic enhancement research poses a number of problems for countries that ban the practice. They could suffer a genetic brain drain as gifted genetic scientists left for more hospitable locations; unless these expatriates were exclusively interested in genetic enhancement, their loss could impede domestic progress in other areas, such as developing gene therapies. Countries that allowed their citizens to become enhanced could put the citizens of other countries at a competitive disadvantage. They also could wage war more effectively. Saddam Hussein is reported, for example, to have hired foreign scientists to help Iraq develop chemical, nuclear, and biological weapons.[40] With the help of a Canadian arms expert, he almost succeeded in constructing a "supergun" that could lob shells into Israel.[41] Enhanced terrorists or armies could be difficult to repulse. Finally, citizens of countries that banned genetic enhancement could obtain enhancements and return home, gaining unfair advantages over their law-abiding brethren. If they enhanced their germ lines, their advantages would be passed on to their offspring, giving rise to the social dislocations described in earlier chapters.

So how could the United States and its allies in the war on genes combat genetic tourism? The U.S. currently spends about $40 billion a year on its efforts to eradicate illegal drugs. About $20 billion is spent by the federal government; the DEA alone has a $1.9 billion budget.[42] It would cost even more to choke off the increased activity represented by shipments of illegal enhancement products into the country. The war on terror has dramatically increased the nation's vigilance at its borders and points of entry; if this war persists, as many believe it must indefinitely, the personnel and electronic surveillance equipment may be in place to intercept illegal enhancement shipments as well as terrorists and their weapons. But it may not be easy to identify illegal enhancement products, at least not those in bulk. Imagine having to search the contents of every person's luggage and carry-on bags for bottles of pills, having to ascertain what the pills are in fact made of, and then, if they have legitimate medical uses, determining the intent to use the pills for illegal purposes.

Even greater difficulties would beset attempts to screen those who traveled abroad and returned with enhanced DNA that had been incorporated into their bodies. "Any DNA to declare?"[43] the customs agent might ask. Returning citizens might be required to submit a DNA sample for rapid analysis; under current law, customs officers can search people entering the country, even in the absence of a warrant or probable cause, so long as they have a reasonable suspicion that the person may have committed an unlawful act.[44] Laws could be passed to allow individuals suspected of harboring genetic enhancements to be quarantined until the testing was completed, in the same way that laws currently permit individuals to be detained for purposes of preventing the introduction and spread of communicable diseases.[45] Perhaps instead of targeting all travelers, those who had spent time in countries where genetic enhancements were known to be available could be singled out for scrutiny. The problem would lie in being able to determine if a person was carrying altered DNA. If the DNA were different from that found naturally, it might be readily detected by fast scanning machines that are likely to be available in the near future. Otherwise, there would need to be some way of comparing the DNA profiles of persons before and after they traveled abroad.

Detecting illegal DNA in persons returning from abroad would be especially difficult in the case of children who had been born abroad after being enhanced during IVF, or in the fetuses of pregnant women. It would be virtually impossible to identify offspring that resulted from passive reproductive enhancement during IVF— the practice of using genetic testing to identify the most desirable embryos for implantation. It would be difficult to detect DNA that had been actively manipulated, since there would be no previous DNA profile to compare it with.

One way to attempt to respond to these detection problems would be to prevent Americans from traveling abroad to obtain enhancements. The U.S. has instituted similar passport controls on travel to "enemy" countries. The list used to include China and Albania; travel to Cuba is still restricted, with some exceptions.[46] Violators can lose their passports, and travel without a passport is punishable by fines and imprisonment.[47]

Besides weeding out people who obtained enhancements abroad, the government could go after those individuals who provided the enhancement services in foreign countries, at least if they were American citizens. The courts have recognized that Congress has the power to enact laws that punish the overseas behavior of U.S. citizens.[48] They can be punished even under laws that do not explicitly extend to foreign activity. In *State* v. *Harvey*, for example, a federal court upheld a conviction for child pornography even though the defendant's actions had occurred in the Philippines.[49] In the same fashion, the law could punish American doctors who operated enhancement clinics in foreign countries. They could lose their medical licenses, and they could be subject to criminal penalties. The problem would lie in bringing them to justice. Physicians who returned to the United States would be subject to punishment for crimes committed overseas, but there would be no effective way to punish individuals who remained abroad. The United States could seek to have the physician extradited for punishment in this country, but this would require the cooperation of the foreign government.[50] It is even difficult for U.S. agents to collect evidence in foreign countries without the help of foreign officials.

The United States also might try to prevent its enhancement technology from being transferred abroad where it could be used by overseas geneticists and entrepreneurs. In the past, the United States has imposed restrictions on the dissemination of technical information that might assist its enemies abroad, or that might accelerate the proliferation of nuclear weapons and other weapons of mass destruction.[51] Dissemination restrictions are common features in government contracts. But much of the scientific work leading up to enhancement breakthroughs would be created with private funding, especially if the government forbids federal funding of enhancement research. The government has attempted to block American scientists from delivering papers at meetings attended by foreigners and from publishing certain works containing sensitive information.

But none of these measures would be successful in the case of information about genetic enhancement technology. Remember that essentially the same techniques employed to fight disease would be used to produce enhancement effects. To stop discussions of enhancement technology, the government would have to restrict disclosure of therapeutic advances. Moreover, the advent of the Internet has made it extremely difficult to interdict the worldwide flow of information. And any government infringements on speech will be met with hostility by academics and perhaps by the courts.

A better approach would be to try to prevent money from being transferred abroad to finance enhancement activities. The government could target funds that might be used to purchase enhancements, to support enhancement research, or to build enhancement clinics. Currency restrictions have long been in place to try to prevent the flow of funds for money laundering, drug trafficking, and terrorism. Anyone who sends more than $10,000 at a time into or out of the United States is required to file a report with the Treasury Department.[52] Federal tax returns require individuals to disclose interests in foreign bank accounts. (This requirement was used to prosecute defendants for shipping arms illegally to the Contra rebels in Nicaragua.)[53] There are specific penalties for conducting foreign transactions with funds from certain unlawful enterprises, which could be amended to include genetic enhancement. Again, though, the problem would lie in detecting illegal financial transactions. And once again, this becomes far more difficult without the cooperation of the countries where the money ends up or through which it travels.

It is clear from the foregoing discussion that it will be hard for the United States to curb international trade in genetic enhancements unilaterally. The war on terror illustrates this if nothing else. Therefore, effective measures will require international action. The United States could attempt to get the United Nations and other multinational organizations to support a global position on genetic enhancement, as they have on issues such as human rights and child abuse.[54]

Work in this direction already has begun in the United States. A group of lawyers and bioethicists met in Boston in September 2001, and issued a call for an international treaty to ban human cloning and germ line genetic engineering. They titled the treaty the "Convention on the Preservation of the Human Species."[55] The preamble specifically cited the concern that "the increased power of genetic science" had the power to "fundamentally diminish humanity" by "intentionally producing an inheritable genetic change," and that "altering fundamental human characteristics even to the extent of possibly producing a new human species or subspecies" could cause the resulting persons "to be treated unequally or deprived of their human rights" or "could easily lead to genocide or slavery." The convention would prohibit human cloning for reproductive purposes and intentional germ line genetic modification.

But more than just an international agreement is necessary for effective controls. Individual nations must ratify the agreement and enact domestic laws to enforce it within their borders. International pressure must be brought to bear on holdout nations, and the United States may have to exert its formidable economic and political power to force recalcitrant countries to toe the line. Under the International Emergency Economic Powers Act, Congress has given the president the authority to impose economic sanctions on countries in order to deal with "any unusual and extraordinary threat . . . to the national security, foreign policy, or economy of the United States."[56] Once the president declares a national emergency, this act allows him to "investigate, regulate, or prohibit" foreign transactions and money transfers.[57] The president has used this authority freely; in 1998, economic sanctions prevented at least three billion people in twenty-nine countries from trading with the United States.[58] Instead of sticks, moreover, the United States could use carrots, providing foreign aid to countries that cooperated with U.S. enhancement policies. To promote worldwide enforcement of its ban, the United States could attempt to get the World Trade Organization (WTO) to adopt its position on enhancements.

The ultimate response that the U.S. could make to nations carrying on an illegal enhancement trade with U.S. citizens would be to use military power. Short of a formal declaration of war, Congress could authorize the president to employ military force against enemy nations, as it did most recently following the World Trade Center attacks.[59] Even without such an authorization, the president has broad war-making powers once he declares that a national emergency has been created by an attack on the nation.[60] While it is difficult to imagine the president invoking this authority to combat the illegal use of enhancements by American citizens, it is not so far-fetched if a potentially hostile foreign country were to employ genetic enhancement to bolster its own military capabilities or to attack the United States.

A glimpse of what the U.S. response might be to such a threat is the war on terror. On October 26, 2001, President Bush signed the USA Patriot Act of 2001. The law increases the money spent on anti-terrorism measures, increases the scope of certain subpoenas, imposes additional trade sanctions, further restricts the movement of money across borders, increases the penalties for money laundering, triples the number of Border Patrol personnel and customs agents, and expands the FBI DNA database.[61]

Declaring a state of emergency has additional repercussions for the government's war on genes besides giving the president greater powers to wage war against other countries. Under the Constitution, the government attains vast powers over its citizens in a state of national emergency; virtually all the liberties protected by the Bill of Rights can be suspended, including rights to procreative autonomy and the right of free speech. In fact, even without a formal declaration of a state of emergency, the courts might view the war on genes as a "compelling state interest" that permits the government to override individual rights. If the threats outlined earlier in this book that genetic enhancement poses to democracy and freedom in fact begin to emerge, the courts may justify all sorts of extraordinary governmental powers, reasoning that they are necessary to preserve the Constitution itself.

Still, a war on genes would exact a great price. Not only would vast resources need to be channeled into the fight, but depriving U.S. citizens of their constitutional rights would not come cheaply. In order to wage this war effectively, the government would have to interfere to an unprecedented extent with private and professional behavior, parental prerogatives, and reproductive decision making. Only the gravest of threats would seem to justify such extreme measures. Despite the warnings in this book, neither the populace nor their elected representatives may view genetic enhancement as such a danger. Even if they did, a war against genetic enhancement ultimately could be self-defeating. Reminiscent of the Vietnam War adage that we had to destroy the village in order to save it, there seems to be little point in imposing a totalitarian regime in order to prevent the erosion of democracy. Unlike a traditional foe, it would likely prove impossible to destroy genetic enhancement once and for all, no matter what steps the United States might take. Given the tremendous advantages that genetic enhancement may afford, the demand—and the willingness of some to fulfill it—will persist. Once democratic institutions were subverted in order to combat the threat, it is difficult to envision how they would ever be restored.

But there is an even more fundamental reason why a total ban on genetic enhancements would be ill-advised. So far we have assumed that the advantages conferred by genetic enhancement would be used by individuals for their personal gain, often at the expense of others who were not so fortunate. Enhanced capabilities could be used to benefit society as well, however. Remember the rescuers in the Introduction. Wouldn't we want them to be as strong and to have as much stamina as possible? How many lives could have been saved at the World Trade Centers if the firefighters had been enhanced to better endure the hazards and pull people from the rubble? If genetic enhancements made people smarter, wouldn't we want our medical researchers to have them so they could more quickly, say, come up with cures for cancer and other dreaded illnesses? If enhancement conferred increased visual and mental acuity and better fine-motor coordination, wouldn't you want it to be available to the pilot of your plane or the driver of your children's school bus? Wouldn't we want to preserve the lives of the soldiers we send into harm's way by making them better fighters?

A ban on all forms of genetic enhancement might deprive society of enormous social benefit. One could go even farther into the future than the foregoing examples. Imagine enhancing people so they could explore the vast regions under the sea, or colonize other planets. In the event of a cosmic catastrophe, genetic enhancement may even be the ticket to the survival of the species. Genetic enhancement could yield incalculable blessings and save countless lives.

The trick, then, is to obtain the benefits without, as much as possible, incurring the harm.

12

Better Solutions

A complete embargo on genetic enhancement is as impractical as it is unenforceable. Researchers will perfect enhancement technologies even if they are barred from receiving federal funds for doing so. Perhaps not as quickly, but eventually. People will illegally enhance themselves and their children. Not everyone, perhaps, and not as many as if enhancements were freely available to anyone with the money to buy them, but enough to transform a ban into a repeat of Prohibition, with the addition of far more government intrusion into people's hitherto private lives and reproductive decision making.

But the real problem with a total ban is not so much that it would be impossible to enforce, but that those who took the risk of thwarting the law would most likely do so for their own benefit. For the most part, the people who became enhanced would not be upstanding public citizens, but criminals. The wealthier criminals, to be sure, but criminals nonetheless. People who felt that their wealth entitled them to purchase optimal characteristics for themselves and their children even though the majority of society had decided that the dangers were too extreme. Selfish people. After all, if you were going to break the law and risk punishment, why would you do it to help others?

So an attempt at a total ban would have the perverse effect of conferring the benefits of enhancement—intelligence, beauty, power—only on the criminal element in our society, while leaving the law-abiding majority at a disadvantage.

There is a way to avoid this, or at least make it less pronounced. Instead of banning genetic enhancement altogether, allow it to be used by the virtuous for virtuous deeds. In other words, don't ban its use, but regulate it. Channel its advantages so they are employed for the good of society.

How can this be accomplished?

ENHANCEMENT LICENSING

There are many instances in which we grant certain individuals extraordinary powers and privileges, but only under certain conditions. Take doctors. They can prescribe potent, often dangerous, medications. They can invade the most intimate parts of the body. They can poke holes in you with needles and, during surgery, lay you open with a knife. They can even kill you—ostensibly with permission, as in Oregon under its physician-assisted suicide law, but sometimes apparently without consent, as when they administer fatal doses of morphine during what is euphemistically called "double-effect euthanasia" and "terminal sedation." They can do all this and not go to jail, much less the gas chamber, because they have a license to practice medicine.

But the license to practice medicine comes with conditions. Physicians must use their powers to heal, not to hurt. In their delivery of services, they must conform to a professional standard of care. They may not extract sexual favors from their patients. They must keep patient confidences secret. They are not allowed to divert narcotics and other dangerous drugs to street use. And so on. If they fail to adhere to these conditions, if they violate the terms of their license, they lose the license, and along with it, their special privileges. They might even go to jail. Or the gas chamber.

The same is true of other professionals. Lawyers, for example, are the only people who can represent others in a court of law.

Besides defendants, jurors, and witnesses, they are the only persons who are entitled to venture beyond the "bar"—the rail that divides the front of a courtroom from the seating area for spectators—and approach the bench. Only lawyers can furnish legal opinions, which can make or break a billion-dollar corporate takeover or the sale of your home. They cannot be compelled to reveal a client's secret even if the secret is that the client has committed a crime. But again, they enjoy these special privileges only so long as they exercise them for the public good. If they aid criminals in the commission of a crime, if they knowingly file lawsuits with no credible foundation, if they lie, cheat, or steal, they lose their licenses.

Licensing is used in a vast number of situations besides regulating the practices of professionals. We license certain individuals to possess or carry guns, to drive cars, taxis, and commercial vehicles. Businesses require licenses. Liquor stores must have them too. Sometimes the license confers a monopoly, an exclusive right to carry on a particular activity or trade. Sometimes it is primarily a means of raising money for the government, like a fishing license. But in all cases, the license is conditioned on obeying certain rules. Taxis must adhere to rate schedules and safety requirements. Fishermen must obey catch limits.

The same approach could be used for genetic enhancement. In the first place, manufacturers and suppliers of enhancement products and services would be required to be licensed. The license would specify the enhancement products they could provide, obligate them to limit sales to licensed purchasers, and impose reporting requirements so that sales could be tracked by the government. Health care professionals would have to be licensed to dispense enhancement drugs or to provide enhancement services. Here again, they would be licensed for specific enhancement technologies, sales would be limited to licensed purchasers, and the purveyors would have to follow strict record-keeping and reporting requirements. (This is similar to the regime employed by the Drug Enforcement Administration to control the production and distribution of controlled substances.) Violators of the licensing terms would lose their licenses and face disqualification from further licensing, fines, and imprisonment.

But the most important part of the licensing program would grant licenses to individuals who wished to become enhanced. These individuals would have to satisfy certain criteria, including having a valid, socially beneficial purpose for becoming enhanced. The purposes for which licenses could be granted would be established by the government, either by the Congress or, more likely, by regulations adopted by the licensing agency that administered the licensing program. This could be the FDA or the DEA, but a better bet would be an independent agency with special expertise in ethics, public policy, and enhancement technology. The prospective licensee also would have to possess whatever background was deemed necessary to help ensure that the enhancement would be used for the purpose for which it was being licensed. A medical researcher, for example, would have to have the training to perform the proposed research. A firefighter would have to have a job as a firefighter. The application would be submitted to the licensing board through a process resembling the one required to obtain a government research grant. This includes submitting a proposal which is reviewed by experts in the field. As in the case of a government grant review, these could be outside experts picked for their expertise in the subjects of the licensing proposals and formed into special "study sections" that would make licensing recommendations to the board. The board would conduct follow-up monitoring to ensure that the license was being properly used for its intended purposes. In the case of government grants, this is accomplished by periodic reports, the production of the proposed "deliverables"—experimental results, for example, in the case of medical researchers—and facility inspections. Government monitoring could be supplemented by appropriate self-regulatory groups. Medical researchers could be monitored not only by the licensing board but by scientific societies like the American Association for the Advancement of Science or the Institute of Medicine of the National Academy of Sciences. Firefighter oversight could be provided by professional firefighters' associations.

The licenses would entitle their holders to obtain enhancement products or services for the purposes specified in their license. Failure to obey the terms of the license would lead to forfeiture of the license and other penalties. These could include confiscating remaining supplies of enhancement drugs, disqualifying the individual from further enhancement licensing, fines and imprisonment. If possible, the enhancement could be reversed biologically.

A licensing scheme clearly would not completely alleviate a black market in genetic enhancements. Individuals who could not satisfy the licensing criteria or who had lost their enhancement licenses might still seek enhancements illegally. There could still be significant genetic tourism. But a licensing system would reduce the demand pressures that would fuel these unlawful enterprises. Health care professionals could provide enhancements legitimately under the licensing system without having to resort to black market sales, thus decreasing the incentive to offer enhancements on the black market. And at least a substantial number of enhanced individuals would be devoting their superior abilities for the welfare of the public rather than merely for their own benefit.

ENHANCEMENT LOTTERY

One of the most pernicious aspects of a free market system in which expensive genetic enhancements would be available only to those who could afford them is the way in which it would undermine equality. As described in Chapter 9, the threat arises not simply because genetic enhancement would exacerbate the actual inequalities between the haves and the have-nots, but because the degree and scope of the advantages it confers would erode the belief in equality of opportunity, the core precept that sustains liberal democratic societies in the face of actual inequality.

The obvious alternative would be for the government rather than the market to distribute genetic enhancements. In Chapter 11, we saw that it would be too expensive for the government to provide genetic enhancements to everyone. But suppose, for the sake of argument, that the government decided to bite the bullet and spend whatever it took to solve the problem. Private purchasing of enhancements would be prohibited, and the government would institute a system for distributing enhancements on a just basis.

It might be argued at the outset that the government has no business providing access to genetic enhancements until it has assured that everyone receives access to therapeutic treatments. But this is not self-evident. Some people may prefer to be enhanced in some respects even while they remain ill or impaired in others. Moreover, the threat from wealth-based access to genetic enhancements may be so great that the government would be asked to step in and provide an alternate means of enhancement distribution even though this means that public resources will be diverted away from therapeutic purposes. But for the sake of argument, let's assume that we have implemented a health system that provides everyone with access to all medically necessary treatments and prevention, and that we are now contemplating how the government can allocate access to genetic enhancements on a just basis. What would that basis be?

How about "on an equal basis"? This has a nice ring. Most Western philosophies agree that equality is a desirable goal. The Declaration of Independence declares it to be a self-evident truth that "all men are created equal." The United Nations puts it more politically correctly in its Declaration of Human Rights, where it says that "[a]ll human beings are created free and equal in dignity and rights." In a passage excerpted by the late Robert Nozick—one of the few well-known philosophers who rejects the value of equality—the great Isaiah Berlin describes the innate appeal of the equality principle:

> No reason need be given for . . . an equal distribution of benefits—for that is "natural"—self-evidently right and just, and needs no justification, since it is in a sense conceived as being self-justified. . . . The assumption is that equality needs no reasons, only inequality does so; that uniformity, regularity, similarity, symmetry . . . need not be specially accounted for, whereas differences, unsystematic behavior, changes in conduct, need explanation and, as a rule, justification. If I have a cake and there are ten persons among whom I wish to divide it, then if I give exactly one-tenth to each, this will not, at any rate automatically, call for justification; whereas if I depart from this principle of equal division, I am expected to produce a special reason. It is some sense of this, however, that makes equality an idea which has never seemed intrinsically eccentric.[1]

Thus far the philosophers—with a few exceptions like Nozick—agree. But they fall to bickering when it comes to deciding what an equal distribution of benefits looks like. Some say that an equal distribution must make everyone equally well-off. Everyone must be given the resources—in our case, the genetic enhancements—that would make them equally happy or equally successful, or whatever the end-state sought. Since people differ in what it takes to make them happy or what they regard as success, this method for achieving equality would end up giving different people different amounts of the resources in question. In terms of how to distribute genetic enhancements equally, this approach, known as "welfare egalitarianism," would produce the result that some people would receive access to more enhancement than others would, because attaining their conception of happiness, or success, or what have you, required greater advantages.

Critics of this view cannot justify its logical extreme, which is referred to as "the problem of expensive tastes." They cannot abide the result that an epicure would be entitled to truffles and champagne while someone who was satisfied with far less would receive, say, only a tuna fish sandwich. These philosophers, who are known as "resource egalitarians," argue that everyone should get an equal share of goods—in our case, genetic enhancements—even if it means some people end up better off in terms of welfare than others.[2] Welfarists object in turn that this would penalize people who had expensive tastes through no fault of their own, perhaps even tastes that they had inherited, and also people who had plain old bad luck, as a result of which their bundle of resources left them in worse shape than someone with the same tastes but with better luck. Other philosophers, mindful of the concerns raised by pure welfare and resource egalitarianism, put forward the notion that everyone deserves a decent minimum of resources or well-being, rather than the same amount of resources or whatever it takes to satisfy their expensive tastes.[3]

At this point, many noted minds turn away in frustration from what might be called substance and focus instead on process. The key, in their view, is not so much what people end up with but how we carry out the distribution. And the trick they devise is to ask us to imagine that we are divvying up the available resources (one of these thinkers, Ronald Dworkin, refers to them as "clamshells") with no idea of who we are, or rather, who we will be after the distribution takes place. We might have expensive tastes, or bad luck, or we might not. The leading exponent of this heuristic device, John Rawls, calls this fictitious predistribution state "the original position" and describes the decision makers as being "behind a veil of ignorance."[4]

All of this makes it very difficult to tell what a government program of equal access to genetic enhancements would look like. Following resource egalitarians, the government could give everyone an equal amount of enhancements, and not allow them to purchase more with their own funds. Assuming that a number of different genetic enhancements were available and we couldn't afford to give all of them to everyone, some method would be needed to decide which enhancements each person would receive. The government might decide that what really mattered was IQ, so that it only distributed cognitive enhancements, but some people might prefer to enhance some other trait.

A solution would be to give everyone a voucher good for the same money's worth of enhancements, and allow them to use it to purchase whatever types of enhancements they preferred. But as welfare egalitarians would point out, this would not result in everyone possessing equal advantages. The people who started out with a more favorable set of natural assets, those who had fared better in the natural genetic lottery, would maintain their superior position. If everyone received an additional twenty IQ points, for example, the people who started out with high IQs would remain smarter than the rest, even though everyone's intelligence had risen twenty points. If genetic enhancement permitted people to achieve a target state, such as a specific IQ level, everyone could be made equal by being given the same IQ. But again, this doesn't allow for people who would prefer to substitute strength or agility or beauty for intelligence as the characteristic to be enhanced.

With no ready method for achieving true equality among individuals, the government might decide that it should aim to provide everyone instead with a decent minimum level of advantages. In that case, it might focus its enhancement distribution program on the least well-off, leaving the rest of society to purchase what it wished and what it could afford on the open market. For example, the government simply could subsidize the purchase of enhancements by the poor. Each poor person, for example, could be given an enhancement voucher, good for a certain dollar value of enhancements. Depending on the going price for enhancements, the amount of the voucher might have to be substantial in order for the poor to obtain any real improvement in their prospects. While this scheme would be much cheaper than providing vouchers to everyone, it could still be prohibitively expensive. If genetic enhancements cost only $10,000 per person—a figure that is likely to be a dramatic underestimate—it would cost more than the entire Medicare budget to provide enhancements just to the thirty-five million people who live below the federal poverty line (which is only approximately $17,000 for a family of four).

Since we are assuming for the sake of argument that the government somehow can find the money for such a lavish new entitlement program, we can put the cost issue aside, but an enhancement subsidy for the poor alone would create another more curious problem: People whose incomes were marginally higher than the threshold for receiving the subsidy but still too low to afford enhancements would now be less well-off than those who had received the subsidy. The new group of least well-offs would now seem to deserve a subsidy of their own. But this would lead to an infinite regression, in which groups leapfrog one another and seek subsidies ad infinitum. The Supreme Court acknowledged this problem in the *Bakke* case when it struck down a University of California affirmative action program that gave preferences to minority applicants:

> Those whose societal injury is thought to exceed some arbitrary level of tolerability then would be entitled to preferential classifications at the expense of individuals belonging to other groups. Those classifications would be free from exacting judicial scrutiny. As these preferences began to have their desired effect, and the consequences of past discrimination were undone, new judicial rankings would be necessary.[5]

The only way to avoid the regression is to declare the disadvantage that triggers the preference or subsidy to be unique; discrimination against African Americans might be characterized in this fashion, for example, because of the unique history of slavery, even though the Supreme Court declined to make this declaration in the *Bakke* case. This would make it necessary to give preferences only to African Americans, and not to other disadvantaged groups. But there is no way to declare one group of poor people to be uniquely different than any other for purposes of giving only one group access to enhancement subsidies.

One way to avoid these difficulties might be to divorce enhancement entitlements from the metric of income or wealth and peg them instead to some notion of a decent minimum quality of life. The government could distribute genetic enhancements, for example, only to those persons who were naturally disadvantaged. From a moral standpoint, this comports with liberal theories of distributive justice like those of John Rawls, who asserts a "principle of redress": "[S]ince inequalities of birth and natural endowment are undeserved, these inequalities are somehow to be compensated."[6] (One might ask why redress should be limited to people who were born with disadvantages, rather than extend it to those who had bad luck later, but for the sake of discussion, let's stick with the more limited proposal.)

But who is naturally disadvantaged? Or, rather, what types of deficits count as disadvantages? We again encounter a problem of subjectivity, a variation on the problem of expensive tastes: You may think that a disadvantage has to affect an important aspect of functionality, like mobility, cognition, reproduction, or being able to work. But I may think that I am terribly disadvantaged because, while I love to sing, I do not have perfect pitch. We could solve part of the problem by giving every qualified individual a voucher and let them enhance whatever trait they wished; you could enhance your ability to work and I could install perfect pitch. But this would still require a determination of what counted as a disadvantage so that a voucher was called for.

The obvious answer would be to compare people to some notion of normalcy. People who were abnormal would be considered disadvantaged and entitled to enhancements. But what is "normal"? We considered this question earlier in Chapter 4 in trying to distinguish gene therapy, which seeks to restore people to a "normal" state of health, from genetic enhancement, which might be said to transcend the bounds of normalcy. In that discussion, we noted that the concept of normalcy was highly arbitrary, often a matter of calculating predetermined statistical deviations from population means. One problem with identifying persons who are abnormal for purposes of subsidizing them is the shifting of the population mean that would result from the enhancement subsidy itself. Suppose, for example, that we accepted the proposition that we should genetically enhance the height of persons who were in the shortest 10 percent of the population. How tall should they become? How about "average height"—the population mean. But the mean height of the population would now shift upwards, and a new group would now occupy the lowest 10 percent. Presumably they would now too be entitled to a subsidy, and so on —the infinite regression problem once again.

Another difficulty is deciding how significant a deviation from population norms would count as a disadvantage. In the case of height, the convention seems to be two standard deviations from the mean, roughly 5 percent of the population. But why not 20 percent, or 2 percent?

Congress encountered these same kinds of problems when it enacted the laws prohibiting discrimination against person with disabilities. What types of impairments count as disabilities, and how severe must the impairment be for the individual to be protected against discrimination? Congress could have taken the position that anyone who could prove that they had been discriminated against because of what the person who discriminated against them regarded as a disability would be entitled to redress, but instead the legislators took a different route. They stipulated that only certain disabilities, which they called "major life activities," would count. They defined these as "caring for oneself, performing manual tasks, walking, seeing, hearing, speaking, breathing, learning, and working."[7] (Curiously, reproduction was not on the list. The Supreme Court, however, has since indicated that it regards reproduction as a major life activity.[8]) Moreover, only persons who suffer a "substantial impairment" of these activities qualify as disabled. Perhaps the same definitions could be used in a program of disability-based entitlements to genetic enhancement.

But remember that at the beginning of this discussion, we assumed that the government already was providing everyone with needed medical care. To the extent that a person had an abnormality and we could fix it, this would come under the rubric of therapy, and the person would already be entitled to it. What we're talking about now is how to distribute access to true enhancements; in other words, to identify individuals who were already "normal" in the sense that they were not ill, malformed, or disabled, or who, while not normal, could not be helped by any existing therapeutic interventions, so that giving them enhancements would be a means of offsetting their plight. The approach used in combating conventional disability discrimination would not work well here.

Moreover, an attempt by the government to identify persons who were naturally disadvantaged for purposes of remediation through genetic enhancement might not sit well with everyone. Persons with disabilities might feel that this would devalue their lives.

Anita Silvers, for example, objects to a fixation on making everyone "normal":

> [O]ur normal modes and levels of functioning are, to an extent that often goes unrecognized, socially relative constructions rather than independent biological facts. Adjusting the environment so anomalous individuals can better flourish can be as compensatory as leveling them. Moreover, enhancing individuals or their groups by magnifying their exemplary performances in some domains can, under some circumstances, sometimes compensate for there being barriers to their performance in other domains of functioning. Wherever strategies that equalize the amount of opportunity individuals have available rather than homogenize the kinds of opportunities they can access are feasible, there is even less reason to suppose that restoring anomalous individuals to normal modes of functioning is a better instrument of justice than enhancing the effectiveness of their anomalous modes.[9]

Radical proponents of this view contend, for example, that persons who are congenitally deaf should be considered a "language minority," and they argue that attempts to cure them by installing devices such as cochlear implants results in "the systematic blocking of [this] language minority from coming into its own and pursuing its way of life."[10] One deaf scholar even calls cochlear implants "[t]he Final Solution."[11]

The reference to the Holocaust recalls a shameful historical episode. As chronicled in Chapter 3, the Nazis' determination to wipe out persons with undesirable genetic pedigrees was the culmination of a eugenics movement that enjoyed widespread currency in the United States and elsewhere. While the government program we are considering is not "negative eugenics," in the sense that it sterilizes or otherwise prevents people with undesirable genes from procreating, nor, strictly speaking, "positive eugenics," which encourages breeding by people with "good" genes, there is something troublingly close to state-sponsored eugenics when the government identifies people who are genetically disadvantaged—read "defective"?—and offers them genetic enhancements either as a fix or to compensate them for their predicament.

Then there is that nagging problem of infinite regression. If we give the disadvantaged genetic enhancements, we make them advantaged, and someone else is now, relative to them, disadvantaged and entitled to enhancements.

Faced with all these conundrums, we might give up trying to make everyone equal or give them an equal amount of enhancements and simply give them an equal opportunity to become enhanced. This brings us full circle, but that's all right, since the whole point is to maintain everyone's faith in equality of opportunity.

An interesting way of accomplishing this would be for the government to establish an enhancement lottery.[12] Winners would be chosen at random and would receive a voucher entitling them to whatever genetic enhancements were legally available. The lottery would be open to everyone, but it would be voluntary. Individuals who objected to genetic enhancement on the basis of morality or religion could decline their winnings. Everyone automatically would be given one chance to win, and no one could purchase tickets. The odds of winning could be adjusted upwards or downwards, and the drawings held at varying intervals, so that a sense of equality of opportunity in obtaining access to genetic enhancements was created and maintained.

States use lotteries to raise money for education and other public purposes. But they serve another function. Those who are less well-off view lotteries as "the only possibility for breaking the cycle of poverty they live in."[13] In short, they create the possibility of a dramatic, sudden increase in upward mobility, and by selecting winners at random, they preserve a sense of equality of opportunity. Of course, current state lotteries are objectionable on a number of grounds. Studies show that tickets are purchased primarily by the poor, who can ill afford the cost. In Maryland, half the tickets are bought by the poorest third of the population, and nationally, one-third of families with less than $10,000 in annual income spend 20 percent of that income on lottery tickets. A lottery is a form of gambling, and like any other form of gambling, is highly addictive for certain individuals. The odds of winning are so low that lotteries are really "the sale of an illusion." As a mechanism for raising state revenue, finally, lotteries are severely regressive.[14]

An enhancement lottery would avoid these problems, however. It would not raise any money because it would not cost anything; tickets would not be for sale, either to the rich or to the poor. Everyone would get one chance.

Lotteries have a long history. They were employed by the ancient Greeks, and to a lesser extent by the Romans, to allocate political office, and Roman emperors gave gifts by lot on festive occasions.[15] Queen Elizabeth I created a lottery in 1566 to obtain money to build a harbor; the British government continued to use lotteries to raise revenue until the late nineteenth century, and started doing so again in 1992. The first state lottery in the United States was begun by New Hampshire in 1964.

Lotteries also have been used in the United States as a fair means of distributing scarce resources. Federal courts have upheld them as a constitutionally acceptable method of allocating scarce public housing and liquor licenses.[16] Closer to home, they have been used to dole out scarce drugs. In 1990, Pennsylvania established a lottery to distribute Clozapine for the treatment of schizophrenics in state mental hospitals.[17] Berlex Labs introduced the genetically engineered drug Betaseron (inteferon beta 1b) in 1993, and used a random process of selection to determine which patients would receive it until 1995, when production capacity became sufficient to meet the demand.[18] A similar process has been used for Invirase, a protease inhibitor used to treat HIV infection, and myotrophin, a drug to treat patients with amyotrophic lateral sclerosis (Lou Gehrig's disease). And a lottery was recognized as the appropriate mechanism in what is probably the most dramatic situation. A court in 1842 was called upon to identify the appropriate manner by which to select who will live and who will be cast overboard from an overcrowded lifeboat. Endorsing the drawing of straws, the court stated, "We can conceive of no mode so consonant both to humanity and to justice, and the occasion, we think, must be peculiar which will dispense with its exercise."[19]

A lottery for genetic enhancements might not be necessary. It would not replace the licensing system described earlier, and instead of a lottery, the government could simply give enhancements for free to anyone willing to use them for the public good. But if it were too expensive for the government to give enhancements to every public-spirited citizen, and if wealthier people were able to purchase enhancements with private funds so long as they agreed to abide by the licensing terms, there might still be a sufficient amount of wealth-based access to enhancement to place the belief in equality of opportunity in jeopardy. The threat to equality would be compounded by the likelihood that, even though they would be licensed to employ their advantages only for the public good, enhanced persons could employ these advantages when they engaged in other activities as well. Later in this chapter we will explore ways of dealing with the potential unfairness that this would create, but for now, it is sufficient to raise the concern that individuals who can afford to become enhanced may be able to secure a disproportionate share of societal benefits despite their licensing restrictions. Since equality of opportunity would be threatened by any of these eventualities, an enhancement lottery might come in handy.

PASSIVE REPRODUCTIVE ENHANCEMENT

Chapter 4 identified a range of genetic enhancement technologies, from somatic drugs to gene insertion and deletion. One set of interventions includes techniques in which DNA is not altered and somatic enhancements like drugs and gene infusions are not used. One such approach entails fertilizing eggs in the laboratory, testing the resulting embryos to identify the ones with the most favorable set of genes, implanting them, and discarding the rest. Another technique consists of giving prospective mates genetic tests to ascertain the likely non-disease characteristics of their offspring, so that the adults can decide if they want to proceed to procreate. Still another approach involves aborting fetuses that do not score well on genetic tests administered in utero for non-disease characteristics. Finally, couples could obtain donor eggs or sperm that had been found superior after being tested for non-disease traits. These techniques do not employ enhancement drugs or DNA manipulation, but rather involve making reproductive choices based on information gleaned from non-therapeutic genetic testing. Hence they are called passive rather than active enhancement.

Passive genetic enhancement raises many of the concerns described in earlier chapters. It entails risk; laboratory testing can commit errors, so that the wrong embryos are implanted or fetuses aborted. Recent studies suggest that the in vitro fertilization (IVF) process itself may cause birth defects and low weight babies.[20] Moreover, as noted in Chapter 5, there currently is little effective government oversight of IVF practices to ensure that they are safe and effective. Since in passive reproductive enhancement using IVF, parents select the genetic makeup of their children from among the options available in the fertilized embryos, the children are deprived of the more random genetic future enjoyed by children who are conceived and brought to term naturally. This leads some to complain that this infringes on the enhanced children's personal autonomy. Passively enhanced children also have a greater chance of inheriting genetic advantages than other children, since passively enhanced embryos or donor gametes are deliberately chosen, or in the case of enhancement-based abortion decisions, permitted to come to term, on account of their desirable genetic profiles. The artificiality of their origins could make their inherited advantages seem less deserved than those of other children, making their subsequent achievements seem inauthentic. IVF and obtaining ideal donor eggs or sperm is also expensive; as mentioned in Chapter 8, IVF costs around $37,000 per live birth, and that doesn't include the costs of genetic testing for enhancement or other purposes. Since assisted reproduction techniques such as IVF and gamete donation are rarely covered by health insurance, and then only to treat infertility, passive reproductive enhancement would be available primarily to those who were relatively well-off and could purchase the necessary services with their own funds. This would make this form of enhancement a threat to notions of equality and equal opportunity. Even nontherapeutic abortions, costing in the range of $250 to $500, may be beyond the budgets of the poor. Finally, as a class, children born passively enhanced would have abilities superior to those of children born naturally. This might make competitions between enhanced and unenhanced children seem unfair. For all of these reasons, from an ethical, legal, and policy standpoint, passive reproductive enhancement might merit the same treatment as any other type of genetic enhancement. If genetic enhancement in general were banned, it might be appropriate to ban passive reproductive enhancement as well. If people had to get a license to obtain genetic enhancements, they might be required to secure a license to passively enhance their children.

Yet in a number of important respects, passive reproductive enhancement differs from other forms of genetic enhancement. In the first place, it consists of making choices only among naturally occurring genetic possibilities. The characteristics of the resulting offspring will remain within familiar limits;[21] there won't be any sudden, dramatic increases in intelligence or strength, no eight-foot-tall teenagers, except for the occasional giant resulting from spontaneous genetic mutation. Over time, the limits of human characteristics will become extended, since by selecting the best of their potential progeny to implant or bring to term, succeeding generations will be operating a sort of crude human animal husbandry program. But due to the natural laws of breeding, these changes will occur slowly and incrementally. As a result, the advantages enjoyed by passively enhanced individuals will be modest compared to the capabilities made possible by active forms of genetic modification.

Moreover, passive reproductive enhancement includes actions and choices that traditionally have been regarded as within the province of individual and parental self-determination. Decisions about whether or not to have a child with another person, whether to carry a fetus to term, or which artificially fertilized eggs to implant in the womb, have all been accorded a high degree of privacy and personal autonomy, even when information from medical and genetic testing is used in making the decision. Government intrusion would be highly controversial and, although not entirely unprecedented, difficult to justify.

The government rarely interferes actively in decisions about whether or not to have a child with another consenting adult. Only incest and polygamy are against the law. Of course, the government did at one time do much more; as described in Chapter 3, the law once permitted the government to forcibly sterilize people it regarded as undesirable so that they would not bear undesirable children, and some states enacted and enforced anti-miscegenation laws to try to prevent interbreeding among different races. But all of these laws have either been repealed or declared unconstitutional. In 1974, a federal judge held that it was illegal to threaten women with the loss of their welfare benefits unless they agreed to be sterilized.[22] In reaching this decision, the judge in that case, Gerhard Gesell, issued the following warning:

> Aided by the growing acceptance of family planning, medical science has steadily improved and diversified the techniques of birth prevention and control. Advancements in artificial insemination and in the understanding of genetic attributes are also affecting the decision to bear children. There are even suggestions in the scientific literature that the sex of children may soon be subject to parental control. . . . Surely the Federal Government must move cautiously in this area, under well-defined policies determined by Congress after full consideration of constitutional and far-reaching social implications. The dividing line between family planning and eugenics is murky.[23]

Legislators in a number of states have proposed linking eligibility for welfare benefits to the use of long-term contraceptives like Norplant, but no such requirement has ever been enacted. The law in some states requires people to get blood tests before they can obtain a marriage license so they can ascertain if they have compatible blood types, but the license is not denied based on the test results. Illinois once required people seeking a marriage license to be tested for HIV infection, but again, a license could still be secured despite a positive test result.

Attempts by the government to restrain another form of passive reproductive enhancement—abortions based on the results of genetic testing for non-disease characteristics—would encounter powerful political resistance from pro-choice groups. At least for the present, the Supreme Court is likely to hold unconstitutional laws that prohibit voluntary abortions during the first trimester of pregnancy, regardless of the reason for the abortion decision. The consensus on freedom of choice is sufficiently resolute that the courts uphold only those state laws that interfere significantly with the right to an abortion during the later stages of pregnancy, even though fetuses are capable of surviving outside the womb long before the point in a pregnancy regarded as viable in early abortion cases.

Passive reproductive enhancement using IVF is also defensible on grounds of reproductive freedom. The practice of IVF itself encounters little opposition, even though as Leon Kass, the conservative head of the president's Bioethics Advisory Commission, admits, most of the highly controversial technologies like embryo research, germ line therapy, and human cloning for reproductive purposes would not be possible without it.[24] Pro-life groups object to the destruction or at least the non-birth of the embryos that are fertilized but not implanted, but focus most of their attention on preserving the embryos rather than halting the practice by which they are created. As mentioned in Chapter 5, there is some growing concern over the health risks stemming from the low birth weight of infants born following IVF, and there is some evidence that IVF babies suffer a higher frequency of birth defects than naturally conceived babies. This might lead to government restrictions on the number of multiple pregnancies that are allowed to come to term following infertility treatment and IVF. The United Kingdom has adopted regulations that limit the number of transferred embryos to three; Belgium has a limit of two.[25] But the reason multiple embryos are implanted following IVF is that it is difficult to tell how many will survive; numerical limits on the resulting births could lead to selective reduction, the practice of aborting some but not all fetuses, which would not sit well with pro-life advocates. Questions also are being raised about the ethics of preimplantation diagnosis for late-onset conditions like Huntington disease and Alzheimer's, but no one seems to be suggesting that the practice be banned. Preconception and preimplantation gender selection triggers considerable ethical debate, but there has been no attempt to prohibit it. In any event, unless you took the position that the harm consisted of discarding the unwanted sperm and embryos, restrictions on gender selection and on preimplantation genetic diagnosis would be extremely difficult to sustain. An objection to IVF based on preventing harm to the children would have little merit.

The discussion in Chapter 11 also showed that government attempts to restrict passive reproductive enhancement might be unconstitutional. Courts that are firmly committed to protecting the right to abortion and the right to decide with whom to have children might be more inclined to uphold government restrictions on passive enhancement techniques that relied on IVF, since judges may feel that the Constitution affords greater protection to traditional forms of reproduction. But there is little doubt that embryo selection for enhancement following IVF would receive far greater protection from the courts than active forms of reproductive genetic enhancement that actually manipulate DNA.

Finally, even if the government were determined to ban passive reproductive enhancement, perhaps after the Constitution had been amended to give it the right to do so, enforcing such a ban would be difficult in practice. It is hard to imagine the government scrutinizing proposed marriages to make sure they weren't occurring for enhancement reasons. The difficulty of distinguishing between therapy and enhancement would make it virtually impossible for the government to prohibit abortions for enhancement purposes; parents always could find some physical or mental deficiency in a fetus to sustain the excuse that the abortion had been for therapeutic rather than enhancement reasons. The same would be true of attempts to ban embryo selection for enhancement purposes as part of the IVF process.

Probably the only way to prevent enhancement-motivated abortions and embryo selection would be simply to ban genetic testing for non-disease characteristics. The prohibition could extend to health care facilities, health care professionals, and laboratories. But this would be tantamount to banning genetic enhancement altogether, an approach that we rejected in Chapter 11. The alternative to a total ban, it will be recalled, would be to license enhancements for the public good, but licensing would not work in the case of passive reproductive enhancement for the simple reason that there is no way to impose the conditions of a license on the resulting offspring. The children might be required to become licensed when they reached the age of majority, but what would we do if they refused? Put them in jail? That would really place the sins of the fathers (and mothers) on the heads of their children. Would we reverse the enhancement process, as was proposed for individuals who violated the licensing law after having been actively enhanced? But since the enhancement in question is passive—that is, only involves a parental choice among naturally occurring patterns of DNA—there would be no way to reverse it biologically. The only remedy would be to handicap the children in some fashion. But this would seem patently unfair while they were children, since they had nothing to do with the circumstances of their births, and waiting to do this until they became adults might be too late to deprive them of the advantages that they had accrued during their youth.

In short, the wisest course seems to be to permit passive reproductive enhancement. Still, this would leave its most effective form—embryo selection for enhancement—beyond the financial means of many couples. Even if private health insurance covered IVF and the necessary genetic testing of embryos for non-disease characteristics, what about people who lack insurance or who are covered under government programs such as Medicaid? Would the government routinely pay for embryo selection for the poor? Unlikely, if only because of its cost. Embryo selection could be included in the package of services given to winners of a genetic lottery, however. But remember that embryo selection involves the deliberate discarding of certain embryos because they aren't "superior" enough. Although it would not have as dramatic an impact on future generations as actively engineering the germ line for enhancement purposes, including embryo selection in the lottery prize could nevertheless be viewed as a program of state-sponsored positive eugenics. This might make the enhancement lottery too costly from an ethical and political standpoint. Only if passive reproductive enhancement were regarded as a serious threat to equality might it be worthwhile trying to incorporate it into the lottery package.

UNFAIRNESS REMEDIES

If neither a ban nor an effective licensing scheme were established to reduce the threats to society from genetic enhancements, people would be able to purchase them on the open market. As we saw in Chapter 9, those who could afford them would gain enormous advantages over those who could not. Even if a lottery were instituted, it wouldn't provide enhancements to everyone. The enhanced would still be way ahead of other people when they competed for the same scarce societal resources. And even a licensing scheme wouldn't completely solve the problem. Licensed individuals would be required to employ their enhancements for the common good, but unless the enhancements were extremely short-acting or temporarily reversible, licensees would remain enhanced at other times, including when they competed with persons who were not enhanced. In all these instances, the advantages enjoyed by the enhanced might make these competitions, and their outcomes, unfair. To prevent this, we need some way to level the playing field.

Traditionally, society does this primarily by what might be called "leveling up." People who are at a disadvantage because of discrimination based on race, for example, are given certain preferences through affirmative action programs. Persons with disabilities also receive special treatment; the law requires employers to hire them even though the employer may have to expend extra resources—referred to as making a "reasonable accommodation"—in order for the employees to be able to perform their jobs.

Interestingly, we tend to use leveling up to provide access to limited resources like jobs or college educations only in two circumstances: when the person who is being given a hand has a disadvantage that makes them fall below societal norms (for example, persons with a physical or mental disability), or when they have a disadvantage like race that is acknowledged to have no true bearing on their ability or performance. (This view of race, of course, was not always prevalent in the United States, even after the Civil War. In *Strauder* v. *Virginia,* for example, the Supreme Court stated that the purpose of the post–Civil War amendments to the Constitution was "to secure to a recently emancipated race . . . all the civil rights that the *superior race* enjoy."[26]) The leveling up that would be required to balance the unfair advantages conferred by genetic enhancement, however, would not resemble either approach. By definition, the unenhanced do not fall *below* societal norms because, if they did, then what we are referring to as enhancements would in actuality be therapeutic or corrective remedies, since they would make those who obtained them, whom we have been calling enhanced, merely "normal." At the same time, unlike persons given preferences on the basis of race, those who were entitled to leveling up because they were not enhanced would in fact be less able.

But there are more significant reasons why leveling up is not the answer in the case of genetic enhancements, reasons that we have previously considered. Leveling up would require us to make sure that everyone had access to enhancements, or at least those persons who were genetically disadvantaged. The former is too expensive, while the latter embroils the government in branding some people as genetically inferior. Moreover, neither would actually solve the problem unless it gave everyone the same or equivalent abilities, which is difficult if not impossible.

If leveling up is not the answer, then the only way to prevent unfairness is to level down. Curiously, there are relatively few instances in which society rectifies the unfairness that results when a person who falls within societal norms competes with someone who is at a comparative advantage. Individual members of the elite do not seem to attract a lot of official leveling-down attention, although the wealthy as a class are subject to some forms of leveling down, such as progressive taxation. The reason may be that, in regard to matters other than wealth, we have too much trouble determining which individuals are superior and by how much. Or it may stem from our desire to secure the benefits that the elite provide to society. Then again, if the elite control society, they can hardly be expected to make things difficult for themselves.

Yet there are some circumstances in which our rules do level down to prevent unfairness. One technique for leveling the playing field is to prohibit the competition from taking place when the result would be unfair. Examples include the ban on the use of performance-enhancing drugs in sports, and the use of weight classes in sports like boxing and wrestling. Not all the examples come from the sports world. The Securities and Exchange Commission, for example, bans "insider trading"—stock transactions that are based on information that is not available to the general public.[27]

Another way to level down is to make the person with the advantage share it with the competition. The law of contracts, for example, requires certain kinds of information to be disclosed before a contract will be enforced at law.[28] This information includes knowledge that the other party has made a mistake about a basic assumption underlying the agreement, for example, that he or she is unaware of a hidden pool of water beneath a piece of residential property.[29]

Still another approach is to handicap the superior party. The rules of horse racing force jockeys who weigh less than others to carry weights in their saddles.[30] Strokes are subtracted from the scores of golfers who aren't as good as the other players.[31]

One type of handicapping occurs when a party is required to obtain the other party's permission to proceed despite a conflict of interest. It is a form of handicapping because it prevents the party with the conflict from taking advantage of the other party without the other party's knowledge. For example, a law firm must notify a client if the firm has been asked to represent someone else whose interests are adverse to the client's, and the firm cannot go ahead and represent both parties without the client's consent. Before enrolling human subjects in a medical experiment, a researcher is supposed to disclose that the researcher has been given stock in the company that makes the drug being tested. Like other disclosure requirements designed to level the playing field, this one applies to information that the other party would not ordinarily possess. But the purpose of the disclosure is slightly different. Rather than placing the parties on an equal footing by depriving the disclosing party of its information advantage, a conflict of interest disclosure sounds a warning to the other party to take steps to protect itself. The potential research subject might more carefully consider the risks and benefits of the experiment to try to determine if participation is in fact safe. Clients with opposing interests might insist that the law firm assign their cases to different lawyers who are on opposite sides of a "Chinese wall," which prevents each of the lawyers from obtaining information detrimental to the other client.

One of the most intriguing approaches to leveling the playing field is to eliminate the arm's length nature of a relationship. The parties are allowed to interact, but the superior party is given the responsibility of looking out for the interests of the weaker party. The legal term for this is a "fiduciary" responsibility. Attorneys are fiduciaries for their clients. So are the trustees of a trust in relation to the beneficiaries, and the directors of a corporation in relation to the stockholders. Physicians owe fiduciary duties to their patients, and parents owe them to their children. None of these relationships is permitted to proceed in the manner of a typical business relationship, which is governed by the general rule of *caveat emptor*, or buyer beware. In a fiduciary relationship, the fiduciary is not allowed to withhold information that the other party might find useful, or to sacrifice the other party's interests for the fiduciary's. Attorneys cannot sacrifice the interests of one client for those of another, even if the second client pays the attorney more. Trustees and corporate directors are not allowed to manipulate trust or corporate assets for their own enrichment. Physicians cannot withhold medically necessary treatments from patients so that a managed care plan will give the physicians a larger bonus at the end of the year. Parents must act in their children's best interest, even when it conflicts with the parents' own interests.

The rationale for treating certain relationships as fiduciary stems in part from the imbalances between the parties. The fiduciary typically has a greater amount of information, or can get information more cheaply. Physicians have gone to medical school, for example, but their patients usually have not. Fiduciary relationships sometimes are imposed when one party can harm another, particularly by misusing information that has been entrusted to them by the other party, such as intimate details of their life or their business operations. In fact, a leading theory of fiduciary relationships says that the purpose of the duty is to *enable* one person to trust the other, so that the entrustor can avoid expending resources on monitoring the other's behavior. Thus, patients are supposed to be able to trust their physicians to act in the patient's best interest. Instead of spending their money on double-checking everything the physician tells them to see if the physician is lying to make more money, the patient instead can spend the money on purchasing more health care.

Fiduciary relationships can be recognized by the courts, under what is known as the common (or "judge-made") law, or they can be created by legislatures or by self-regulatory groups. The fiduciary nature of the attorney-client relationship is a creature of the common law. However, the National Association of Securities Dealers has a "conduct rule" that requires a stockbroker making a recommendation to a customer to buy stock to have "reasonable grounds for believing that the recommendation is suitable for such customer upon the basis of the facts, if any, disclosed by such customer as to his other security holdings and his financial situation and needs."[32]

All of these techniques for leveling down could be employed to level the playing field when enhanced individuals compete with persons who are not enhanced. Certain kinds of interactions, like sports contests, could be prohibited outright if they conferred few benefits on society, or if the societal benefits were outweighed by the unfairness or risk of harm to the unenhanced competitor. Enhanced individuals might be required to share their advantages with the unenhanced, revealing information or giving the unenhanced a hand at accomplishing the task. The enhanced could be placed at a disadvantage in certain competitive situations, like being given less time to complete a race or a harder question to answer on an examination. In a sort of reverse affirmative action program, fewer places could be open to them in a business or college. At a minimum, the enhanced could be required to disclose that they are enhanced. And certain interactions between enhanced and unenhanced individuals seem ideal candidates to be treated as fiduciary relationships. Persons who were enhanced could be required to act in the best interests of the unenhanced, or even be held responsible for their welfare, a sort of genetic *noblesse oblige*.

Leveling down could be imposed as a condition for obtaining an enhancement license. For example, licensed persons could be required to disclose that they are enhanced as a condition for retaining their licenses. Imposing a fiduciary obligation on enhanced individuals as a licensing requirement seems especially apt, since fiduciary obligations are imposed as a licensing condition on lawyers, doctors, accountants, psychologists, and other professionals whose licenses give them special powers and privileges.

Yet we permit many interactions to take place without leveling the playing field, despite the potential for unfairness. Many athletic competitions force older athletes to compete against younger ones instead of establishing age categories. One sportswriter describes baseball, for example, as a "young man's game," in terms of how difficult it is for older players to remain on teams.[33] In football, a player with a slight build takes the field at his peril. No attempt to level the playing field is made in basketball; shorter players are not allowed to shoot, for example, from stepladders. Nor are there any professional basketball leagues for players of normal height (although someone apparently once tried to start one).[34] Competitive chess only began banning the use of cognitive enhancers like nicotine and stimulants in 1999.[35]

The reason we ignore certain advantages in these situations is probably due largely to historical accident. But leveling down also is controversial. A leading advocate of meritocracy, which holds that people should be rewarded based strictly on their abilities, observes, for example:

> [I]n efforts to minimize the differences in performance, we can detect not only the hand of the generous person who honestly regrets that some must lose the foot race but the hand of the envious ones who resent achievement, detest superiority in others, and will punish eminence at every opportunity. These latter are the ones Henri Becque had in mind when he said, "The defect of equality is that we only desire it with our superiors."[36]

And in the short story *Harrison Bergeron,* Kurt Vonnegut includes the following acerbic description of a future in which people who are more talented than others are handicapped:

> The year was 2081, and everybody was finally equal. They weren't only equal before God and the law. They were equal every which way. Nobody was smarter than anybody else. Nobody was stronger or quicker than anybody else. All this equality was due to the 211th, 212th, and 213th Amendments to the Constitution, and to the unceasing vigilance of the United States Handicapper General.[37]

Leveling down the genetically enhanced would not seem appropriate, for example, if they were not competing for scarce resources, if the competition were not a winner-take-all or "zero-sum" game, or if there was no way to shift to the unenhanced the advantages stripped from the enhanced. Philosopher Thomas Nagel, for example, objects to egalitarians who "would permit haute cuisine, haute couture, and exquisite houses to disappear just because not everyone can have them."[38]

Leveling down also would be a mistake if it deprived society of a substantial amount of benefit that could be obtained no other way. For example, because we want to obtain as quickly as possible the scientific knowledge that will enable us to cure diseases, we would allow enhanced physicians and scientists to grab the bulk of the research grants from the NIH rather than reserve the money for the unenhanced. Similarly, we would not hobble enhanced soldiers or policemen merely to make things more fair for the enemy or the thief. And when enhancement can reduce the risk of harm to third parties, we should not worry too much that only enhanced people can become airplane pilots or automobile mechanics.

Leveling down wouldn't be easy. We would need to be able to identify who was enhanced, part of which would entail the often difficult task of distinguishing between enhancement and therapy. To avoid discouraging effort, we would need to distinguish between talents instilled by genetic enhancement and those attained by hard work. We would need to fine-tune the leveling process or we would be ignoring the fact that some unenhanced individuals may have other significant advantages, like wealth or special knowledge, that may help offset the effects of enhancement. And let's not forget that leveling would be intrusive and administratively costly.

But we do level down despite these concerns when it seems necessary. If necessary, we even do so without making fine adjustments. Progressive taxation deprives people of wealth that has been earned. Weight classes in sports are maintained regardless of whether an athlete's size is natural or the product of strenuous diet and exercise. Fiduciary rules require information to be divulged even though it may have been obtained at great effort and expense. If necessary, we put up with the intrusion and the cost of leveling the playing field. Athletes are forced to submit to physical examinations and to provide samples of body fluids for testing, often under conditions that deny them minimal privacy. If they don't like it, we tell them, don't play sports.

Reducing the unfairness of genetic enhancement is one of these situations. If you don't like being leveled to promote competitive fairness, don't become enhanced.

GERM LINE ENHANCEMENT

The previous chapter laid out the problems raised in trying to ban genetic enhancement. A ban would stimulate the creation of black markets. People would just get their enhancements abroad. It would cost society dearly, both in terms of the costs of a war on genes and the loss to society of the fruits of enhanced talent.

Yet there is one type of genetic enhancement that must be banned nonetheless: active germ line enhancement.

Germ line enhancement alters the genome to produce enhancement effects, and, in contrast to somatic enhancement, does so in such a way that the genetic alterations are passed on to future generations. For this to happen, the germ line alteration must reach and be incorporated into the germ cells—the cells that will become eggs or sperm. For that to take place, the alteration must be made before the germ cells are formed—at early stages of embryonic or fetal development.

The dangers from germ line genetic enhancement were described in previous chapters. Briefly, if the enhancements are broad and powerful enough, they will give those who possess them such tremendous social advantages that they will come to dominate society. Unlike somatic enhancement, the advantages conferred by germ line enhancement will be inherited by the children of the enhanced. Succeeding generations will capture all that is worth having. They will form a new aristocracy, a genobility, that will destroy the belief in equality of opportunity and the foundations of democracy. Society will degenerate into a caste system, and then to slavery or chaos. If the germ line changes are radical enough, the new lineage will begin to transcend the limits of the human race. Ultimately, more than one sentient species will occupy the planet, and to those of the unenhanced who still survive, the enhanced will seem like gods.

None of the restrictive techniques described so far in this chapter could prevent this from happening. The licensing scheme would not ensure that inherited enhancement advantages would be used only for public benefit, since there would be no way to bind the offspring to the conditions of the license. Including access to germ line enhancement among the lottery prizes would not eliminate the genobility, but would only swell its ranks. Leveling the playing field would not prevent germ line engineering, and besides, it would only be a matter of time before the enhanced aristocracy accumulated sufficient power to cast off the leveling restrictions.

The only way to prevent this nightmarish future is to prevent germ line enhancement from taking place. To begin with, there should be no federal funding for germ line enhancement research. As noted earlier, though, the technologies that would be used for enhancement, including germ line enhancement, will be the same as those employed for therapeutic purposes: sequencing, but in this case sequencing non-disease genes; incorporating detectors for these genes in DNA chips used in high-speed therapeutic testing; producing enhancement drugs using recombinant DNA; developing more sophisticated drug delivery systems; employing genetic testing for non-disease characteristics in preconception couples screening, embryo selection as part of IVF, and fetal testing; and, finally, transfer technology that edits DNA itself. Even without federal funding for explicit enhancement research, these technologies will get a big push from the NIH. Adapting them for enhancement purposes would likely be a snap. You might think this could be nipped in the bud by withholding federal funds for the sequencing of non-disease genes. Without knowing where these genes are, how they interact with other genes and with the environment, and the sequence of their nucleotide bases, none of the other types of enhancements will be possible, including germ line enhancement. But except for grant applications that expressly propose to perform enhancement research, it would be virtually impossible to block this type of basic scientific investigation. Remember there is no bright line between enhancement and therapy. Genes that might be manipulated to produce enhanced intelligence, for example, are likely to be the same genes that cause mental retardation or the cognitive impairments of old age. We will find them to cure disease, and they will then be available for enhancement use. In short, a ban on government-funded germ line enhancement research will not halt the research. But it may slow it down.

The commercial value of highly effective germ line enhancement will make enhancement research an attractive investment for the private sector, so that germ line technology would be perfected even without federal research funds. Therefore, private research on germ line enhancement must be banned as well, even though, for the reasons just elaborated, such a ban will have only a limited effect. In addition, the patent laws should be changed to bar the issuance of patents for germ line enhancement.

Various methods can be used to enforce a research ban. Companies and researchers that conduct germ line enhancement research could be disqualified from receiving federal funding for other research, and from having their therapeutic and somatic enhancement products approved by the FDA. Hospitals and clinics that carried out germ line enhancement testing on their patients could be barred from receiving Medicare and Medicaid reimbursement. Since this might have little impact on infertility clinics, which receive most of their money from their patients out of pocket, federal or state law could require these clinics to be licensed, and the licenses could be forfeited if the clinics were caught performing germ line enhancement research. State law could revoke the licenses of physician researchers as well.

Research must be prohibited on animals as well as on humans.

In the whole history of genetic science, this has only happened once. At the very beginning of the modern revolution in genetic engineering, because of concerns over risks to the environment, scientists willingly accepted for a brief time a moratorium on recombinant DNA research, which at the time was being carried out only on single-cell organisms. Even when we propose to stop certain research on humans, it continues in creatures lower down on the evolutionary scale. While President Bush and his allies limit embryo and stem cell research in this country and push for a ban on human reproductive cloning, animal research labs continue to pump out Dolly the sheep, "cc" (for "carbon copy") the cloned kitty, and Alba the transgenic, glow-in-the-dark bunny whose DNA was combined with that of a fluorescent jellyfish for the sake of art.[39]

Traditionally, animal research has only been regulated to avoid cruelty to the experimental subjects. While Humane Society representatives may object to germ line enhancement experiments in animals for this reason, that would not be the main purpose of a ban on animal enhancement research. Nor would it be primarily aimed at slowing down the preliminary research in animals that precedes trials in humans, although it would have that salutary effect. The real reason why germ line enhancement research in animals must be prohibited—and another glimpse of why the prospect of germ line enhancement is so frightening—is that it could be another route to creating a rival intelligent species.

This leads to another chilling observation. One of the objections to human cloning and genetic engineering sometimes is that human DNA could be manipulated to create armies of semi-human soldier-drones to fight future wars, creatures that were abnormally strong and aggressive, with heightened pain tolerance and a reduced sense of self-awareness so they would blindly accept suicide missions. But why start with human DNA? Why dumb down people so they will carry out risky or hare-brained military schemes? Why not dumb up apes?

Returning to the details of a ban on germ line genetic enhancement research in humans, the prohibition needs to extend beyond research. Purchasing or providing germ line enhancement, or transporting it across state lines for purposes of sale or distribution, must be made illegal, and anyone caught doing it must be severely punished. Mere "possession" of a germ line enhancement—the state of having or inheriting altered DNA in one's germ cells—should not be a crime because this would punish children who have no control over the actions of the parents or prior ancestors who ordered the engineering. For purposes of the ban, germ line enhancement should be defined as the insertion or deletion of DNA for enhancement purposes where the effect is intended or may reasonably be expected to affect reproductive cells (eggs, sperm).

The power of the government to impose the ban would derive from the government's constitutional authority to promote the general welfare and to protect democratic processes of government. In the parlance of constitutional lawyers, as noted in Chapter 11, this is a "compelling state interest," which means that the scope of the legitimate exercise of this government power is the broadest possible. As a consequence, many of the government's actions required to enforce the ban that would be unconstitutional if employed for other purposes, such as certain searches and seizures, would be permissible.

There is one further sanction that can be imposed on persons whose germ line becomes enhanced. They can be sterilized, so that they cannot pass the enhancement modification to their children. This is a drastic and heinous step and should only be considered as a last resort, if all other efforts fail to limit access to the technology. Would such a measure be constitutional? The state would assert a compelling state interest, which may be able to override the enhanced individual's fundamental interest in procreation. The government also would have to overcome the equal protection barrier erected in *Skinner* v. *Oklahoma,* discussed in Chapter 11. The courts would have to find that the reach of the sterilization law was neither too narrow nor too broad, so that it treated individuals in like circumstances the same.

The ban on germ line enhancement should not be total, however. For the reasons given earlier in this chapter, it is still advisable to permit passive reproductive enhancement to take place, and this will have a small and indirect but definite germ line effect. Embryos that were discarded and fetuses that were aborted for enhancement reasons would not have an opportunity to reproduce, and therefore would not contribute their mix of genes to the gene pool. Donor gametes and embryos brought to term that had been selected for their superior characteristics, on the other hand, would pass their genes on to their offspring. But the resulting changes should be slow and modest enough to avoid the major problems of active germ line enhancement. And these more traditional forms of reproductive decision making are entitled to greater constitutional respect.

Another issue involves germ line gene therapy. As noted earlier in Chapter 4, germ line therapy is itself controversial. There are fears that we don't know enough about genetic science to ensure that germ line therapies are safe. They may turn out to have terrible, unpredictable side effects that may not be apparent for many years or even for generations. On the other hand, some individuals whose lives have been sorely tried by inherited disease argue that germ line alterations are the only sure way to eradicate these diseases completely. The debate undoubtedly will rage on as research on animals perfects the underlying technical skills that would be employed in humans.

If germ line gene therapy ultimately is permitted, then in order to continue a ban on germ line enhancement, it will be important to be able to distinguish between them. Much earlier discussion has pointed to how difficult this might be. After all, this was one of the reasons why the idea of a total ban on genetic enhancement was rejected at the beginning of this chapter. But neither a ban on germ line enhancement nor a workable licensing scheme for somatic enhancement will be feasible unless a method is found to separate therapy from enhancement. Researchers, health care professionals, and potential licensees will need to know what is permissible and the police will need to know what is illegal. This will not be easy, and some of the answers will only be available following case-by-case decisions by the NIH, licensing boards, law enforcement personnel, and the courts. But a concerted and sustained effort must be made to define the boundary. Some of the techniques that may be necessary will be discussed in the next chapter.

Distinguishing between therapy and enhancement will not be the only problem in enforcing a ban on germ line genetic enhancement. All of the enforcement challenges mentioned in Chapter 11 would need to be confronted, the difficulty of controlling off-label uses of therapeutic drugs and offshore access to enhancements; the interference with personal liberty; the cost. But the threat from germ line enhancement is serious and profound enough to warrant accepting the challenges.

Moreover, one major reason why a total ban on genetic enhancement would be unwise should not deter banning germ line enhancement: Banning germ line enhancement should not cause a significant loss of societal benefit. True, just as gene therapy may be most effective if carried out an early stage of human development, there may be some capabilities or ability levels that may only be achieved by enhancing embryos or early stage fetuses, which inadvertently, at least, would affect their germ cells. But the enhancement effects made possible by somatic interventions ought to be more than enough to permit individuals to confer significant benefits on society without the dangerous potential impact of germ line enhancement engineering. After all, the societal benefits we want to obtain from enhancement do not include the destruction of democracy and the enslavement of the species.

CONTROLLING FOREIGN ACTIVITIES

As Chapter 11 predicted, restricting access to genetic enhancements in the United States would drive people overseas. There they would purchase enhancement drugs to bring with them when they return, or they'd simply purchase the products over the Internet and have them shipped. For enhancements that require more elaborate medical interventions, like IVF, they would go abroad and return after they receive the procedure or, in the case of enhancement during reproduction, after becoming pregnant or after delivering the child, just like in the old days when women who had conceived out of wedlock would disappear from society's view for a discreet period. Overseas, people seeking genetic enhancement would find pharmacies eager to sell and clinics ready to serve. No licenses would be required, or they would be easily obtained on the black market. Whole industries devoted to developing enhancement technologies would take root in hospitable countries, where they would be built with foreign investment and staffed by foreign scientists eager to make money by pursuing their interests in forbidden areas of research. All of this would happen unless we made controlling foreign sources of enhancement interventions one of our foreign policy priorities.

The United States could take a number of unilateral actions to restrict overseas access to genetic enhancements. The government agencies engaged in waging the war on drugs can add illegal enhancements to their lists of controlled substances. Customs and Coast Guard officials could be on the lookout for enhancements when they inspect suspicious shipments, and could subject returning citizens to searches of their belongings, and their bodies, including their DNA. Congress could make it illegal to travel abroad for purposes of obtaining enhancements without a license, just as it has criminalized pornographic activities engaged in by U.S. citizens with children in foreign countries. Corporations chartered or doing business in the United States could be barred from operating or investing in foreign enhancement industries. Banks and other financial institutions could be prohibited from transferring funds to Americans in countries where enhancements were freely available.

But as noted in Chapter 11, effective measures to control the overseas enhancement trade would require the cooperation of other countries. Without this, the United States would be unable, short of waging war, to shut down foreign enhancement operations or to capture and extradite American citizens who break U.S. enhancement laws. Only if other countries enacted and enforced domestic laws similar to those in the U.S. would the U.S. be able to mount a successful campaign against illegal genetic enhancement. To get other countries to go along, the United States would need to use all its tools of political and economic persuasion. It would have to convince international and regional organizations and individual countries to sign and ratify treaties regulating the enhancement trade. As a result, the U.S. may have to delegate a substantial portion of enforcement responsibility to the United Nations or other international bodies. Recalcitrant nations would need to be bullied with trade sanctions or bribed with foreign aid. Bi- and multilateral trade negotiations would need to include discussions of the need to regulate genetic enhancement. The World Bank and the International Monetary Fund would need to play prominent roles.

If all this fails, the final option is the use of force. This may seem far-fetched, but it cannot be ruled out if genetic enhancement overseas is perceived as a sufficiently serious threat to U.S. interests. Consider our reaction if a hostile nation like Iraq was vigorously pursuing enhancement research, with the aim of creating enhanced soldiery or of destabilizing the United States by making illegal enhancements available to its citizens. In that case, genetic enhancement might be regarded as a biological weapon, similar, say, to anthrax. The United States might well decide that it must eliminate an unfriendly nation's ability to produce enhancement technologies, with military means if necessary.

It is difficult to imagine now that controlling genetic enhancements abroad will ever come to occupy such a prominent place among the objectives of American foreign policy. At present, there seem to be many higher priorities, like winning the war on terror. But the threats from unrestricted access to genetic enhancements cannot be ignored indefinitely. Eventually we will need to act, both domestically and in the global community.

13

Spotting Enhancement

The approaches described in the preceding chapter for dealing with the societal threats from genetic enhancement depend for their effectiveness on the ability to identify enhanced individuals. Some of the techniques for enforcing a total ban on germ line genetic enhancement, such as reversing the genetic engineering or, as a last resort, preventing the enhanced from procreating, depend on being able to detect persons whose germ cells contain altered DNA. A licensing program for somatic enhancements must be able to catch unlicensed users as well as those who illegally provide them with enhancement products and services. Leveling the playing field to reduce unfairness cannot take place unless we can discern who is and who is not enhanced.

A similar problem plagues athletic organizations that prohibit the use of performance-enhancing drugs in sports competitions. As described in Chapter 7, the Olympics goes to enormous lengths to catch violators, constantly increasing its testing requirements, devising more sensitive tests for existing substances, and designing new tests for the latest drugs. Yet Olympic officials are constantly challenged by athletes and coaches, aided by unscrupulous scientists, who strive to stay one step ahead of the testing regimens.

The 1997 film *Gattaca* depicts a similar testing program aimed at genetic enhancement in the future.[1] It opens with white-collar workers entering their office building and pausing to have some cells scraped off their fingertips so that their DNA can be tested. Later, security guards on the lookout for illegal employees vacuum up pieces of hair and fingernail clippings from computer keyboards and subject them to testing. DNA testing is constant and ubiquitous, the overall atmosphere totalitarian and oppressive.

Is this the only alternative to the destructive forces of genetic enhancement? Is this our future?

In some respects, unfortunately, the answer is probably yes. There is no way to implement a licensing program or a ban on germ line engineering, to take steps to reduce unfairness, without some sort of widespread surveillance system, and this would require genetic testing to be conducted with a frequency resembling that used in competitive sports. There might need to be random DNA screening to catch people with germ line enhancements. Automatic examination of persons returning from foreign vacations and business trips. Mandatory testing before and during whatever competitive events require steps to be taken to level the playing field. Students required to give DNA samples being taking college entrance exams or finals.

This is not a pretty picture, and it gets worse. Some physical enhancements, like an extreme increase in height, might be discernible to the naked eye. But many enhancements, like those that improve cognitive functioning, would be much more difficult to detect. How would we tell whether someone's DNA had been enhanced? How could we determine if a child had inherited an illegal germ line enhancement?

It may be possible to "tag" DNA that has been intentionally manipulated. Agricultural biotechnology companies have developed methods for inserting non-functioning nucleotide sequences into genetically modified crop genomes so they can tell if farmers have used their patented seeds and make sure the farmers have paid for the privilege.[2] Similarly, it might be possible to tag human DNA inserted to achieve an enhancement effect, and perhaps even to leave a non-functional nucleotide footprint when DNA is deleted for enhancement purposes. Conceivably, such a tracer could be incorporated into the recombination process used to manufacture somatic enhancement products. High-speed sequencing machines could then detect the telltale nucleotides.

If this type of tag were to be found only in legally obtained enhancements, then genetic testing could reveal whether or not an enhancement that someone had employed was contraband. The DNA tag would resemble the watermarks and other devices that are used on paper money to prevent counterfeiting. The tag even could be added at the point that licensed individuals received their enhancements, so that the sequence could contain a unique identifier, such as the license number. (This is similar to the system used by alcoholic beverage control agencies years ago, in which the last four digits of a store's liquor license were hand-stamped onto a tax stamp that was glued to every bottle of liquor; as a result, someone with a list of license numbers could always tell at what store a bottle had been purchased.)

The question is: How difficult it would be for rogue scientists to duplicate the DNA tag and add it to illegal enhancements? The technology may defeat the unsophisticated, but at the price that people will be willing to pay for high-caliber enhancement products and services, it is hard to believe that clever geneticists could not replicate the sequence. At that point, how would we be able to detect persons who have obtained enhancements on the black market? How would we tell if someone had inherited enhanced DNA illegally from their parents? Or that they had returned from abroad after an infusion of altered, untraceable DNA? Or even that they had merely swallowed an enhancement pill without being licensed?

The process for identifying people who illegally use enhancement drugs made with recombinant DNA would resemble the programs used to detect performance-enhancing drugs in sports. But to catch people who illegally altered their DNA or the DNA of their children, even more elaborate measures would be necessary. We would need some way to compare their DNA to some prior, original state. In order to maintain an effective testing program, we would need a massive, centralized DNA database.

Everyone's genetic profile would be entered at birth. When they were tested later to ascertain if they were illegally using unlicensed enhancements, the test results could be compared with the sample obtained at birth; illegal gene insertion or deletion would show up because their nucleotide sequences would have changed. In addition, we would need software programs that, by analyzing the DNA of its parents, could predict what a child's natural genetic profile would look like with sufficient accuracy that we could identify children whose DNA had been tampered with illegally. Some of these measures would require an enormous increase in our DNA testing capabilities. All of them are intrusive and would create threats of discrimination and other forms of stigmatization if sensitive genetic information got into the wrong hands.

But it is likely that we will see universal DNA testing and population-wide data banks long before we need them to help enforce restrictions on genetic enhancement. The growing field of pharmacogenetics aims to tailor medical treatment more efficiently and with fewer side effects based on patients' DNA profiles. Increasingly, DNA sampling and testing will be a routine feature of medical examinations. The desire to use the highly effective tools of DNA identification to fight crime and to win the war on terror create powerful incentives to implement population-wide DNA screening and store the results in a single government database. The obvious approach is to perform this screening at birth simply by adding DNA testing to current newborn screening programs. Whether we like them or not, these measures are likely to become part of our lives in the not-too-distant future.

This isn't to say that we should be cavalier about the interference in our lives that will result from efforts to regulate genetic enhancement. Nor that we should fail to take whatever steps we can to protect personal privacy and prevent the inappropriate use of genetic information.

But this is the price that we will have to pay to preserve as much of our way of life as genetic enhancement will permit.

And as you perhaps contemplate the bleak image of the future portrayed on the screen by the movie *Gattaca*, bear one other thing in mind. The testing that occurs throughout that film is designed to detect unenhanced persons, like the hero, so they can be barred from the best jobs and other societal rewards. The genobility is using genetic profiling to maintain its privileged status. The genetic testing program I am proposing, by helping to enforce a ban on germ line genetic engineering, by supporting a licensing system, and by endeavoring to alleviate unfairness, would be instituted for exactly the opposite purpose.

14

Conclusion

Orwellian scenes of government surveillance. Drastic curtailment of reproductive liberty. Class warfare. Covert action against other nations. Monstrous creatures asserting supremacy over other human species.

Is this vision of the future inescapable? We surely hope not. And just possibly, hope will be enough. All of this may turn out to be an overinflated science fiction story, a child's nightmare. Genetic enhancements may never be developed. If they are, they might not work. If they do, society somehow might learn to accommodate them without major dislocation.

If you listen to some genetic scientists, you will hear a lot of this talk. Let's hope they're right. But there is a subtext beneath some of their comments dismissing genetic enhancement that suggests a disturbing reason why their views might be somewhat skewed. In an article in the leading journal *Science,* for example, Jon W. Gordon rejects the prospects for genetic enhancement, stating at one point:

Given the inherent limitations of the gene transfer approach to enhancement, discussion of extending such procedures to humans is scientifically unjustified. We clearly do not yet understand how to accomplish controlled genetic modification of even simple phenotypes. Where more complex traits such as intelligence are concerned, we have no idea what to do, and in fact we may never be able to use gene transfer for enhancement of such phenotypes.[1]

He then proceeds to ask: "If we accept the notion that genetic enhancement is not practicable in the near future, what policies should we develop concerning the use of such technology?" After explaining why attempting gene transfers for enhancement purposes would be unethical, he concludes with the following caution:

Finally, and perhaps most important, broad legal restrictions incur the risk of limiting invaluable research. . . . If, as a society, we feel compelled to make a statement against genetic enhancement, we need not enact anticipatory legislation. Instead we can evaluate such manipulations as we would any other invasive clinical procedure . . . [and] we will currently reject the procedure on all counts as medically unethical.

Fear of genetic manipulation may encourage proposals to limit basic investigations that might ultimately lead to effective human gene transfer. History has shown that effort is far better spent in preparing society to cope with scientific advances than in attempting to restrict basic research. Gene transfer studies may never lead to successful genetic enhancement, but they are certain to provide new treatment and prevention strategies for a variety of devastating diseases.

In short, Gordon is worried that, as a result of fears over the threat of genetic enhancement, NIH funding for basic genetic research will be cut off and other restrictions might be added. The question is whether these fears lead him and other scientists to overstate the technical difficulties and understate the true potential when they downplay the prospects for genetic enhancement. When they say that genetic enhancement will not be a reality in the foreseeable future, are they being straightforward, or are they primarily protecting their grants?

The fact is that we don't know. And the real danger is that we may not know until it is too late to do anything about it.

Our predicament is worsened by the fact that the actions we would have to take to reduce the threats from genetic enhancement are extremely unpalatable. No one wants to impose a *Gattaca*-like testing regime. No one wants to interfere with reproductive freedom without a very good reason. Restrictions on genetic research cannot be undertaken lightly. It will be very costly to regulate genetic enhancement in a sufficiently fine-tuned fashion that we obtain societal benefits with a minimum of risk to the foundations of democracy. In the meantime, other priorities preoccupy us: The war on terror, racism, the economy. Almost certainly, we will not implement the necessary measures to deal with the threat of genetic enhancement until we have incontrovertible evidence that they are warranted. We will wait to close the barn doors until we see the horses running. And we will hope we are in time.

Fortunately, some of the steps that must be taken do not have to be taken right away. There is no reason to establish a licensing program until more genetic enhancements become available. A lottery is not yet necessary, since there are still few enhancements that only the wealthy can purchase. But some steps must be put in place now, or in the near future. It is not too soon for Congress to begin considering a ban on deliberate enhancement research, whether by publicly or privately funded entities, and start figuring out how to design it so that it interferes as little as possible with basic science research and with the development of effective gene therapies. The conversation among nations about how to control genetic enhancement globally must start immediately.

There is more that we can do now. For one thing, we can urge the NIH to revitalize the Recombinant DNA Advisory Committee (the "RAC") and restore its concurrent jurisdiction with the FDA over genetic enhancement research and development. The RAC has the advantage over the FDA in that it is designed to bring together a multidisciplinary group of experts to review the ethical, legal, and social implications of new genetic technologies. These dimensions are far afield from the FDA's traditional oversight of drugs, medical devices, and biologics. If the NIH is unwilling or unable to revive the RAC, then at the least the FDA ought to adopt its composition as a model for its own deliberations on enhancement research protocols and product licensing.

Another thing we can do now is begin watching carefully for signs that the pace of enhancement development is picking up. The canary in the mine, so to speak, might be the first truly effective drugs to combat the cognitive impairments of Alzheimer's disease and old age. It is almost certain that these drugs will be developed with recombinant DNA or key insights from genetic science, qualifying them as genetic enhancements. They will present the first great challenges to society from this emerging technology, in terms of the safety and effectiveness of their off-label enhancement use; the degree to which individuals feel compelled to take them to succeed; the way they call into question the authenticity of achievements accomplished with their aid; how unfair they make competitions for scarce resources; and, if they are very expensive, how seriously they undermine equality of opportunity.

In the meantime, we can ready our responses. Think carefully and debate publicly the alternatives. Design experiments and stand by to try them out to see what works best.

And continue to hope for the best. Even if we put all these programs in place, there can be no guarantee that they will work. For that to happen, we will not only need wisdom and courage, but a great deal of luck.

Notes

INTRODUCTION

1. Statement of Theodore Friedman, First Gene Therapy Policy Conference, National Institutes of Health, September 11, 1997.

1. AN ANNOUNCEMENT AT THE WHITE HOUSE

1. Much of the story told here is based on an article by Nicholas Wade, "Genetic Code of Human Life Is Cracked by Scientists," *New York Times,* June 27, 2000.

2. Skeptical accounts are reported, respectively, in Ken Ringle, "Violent Nature: Behavioral Scientists Consider Why Some Kids Reach for a Gun," *Washington Post,* April 24, 1999; "Virus 'Footprint' Found in Schizophrenics: Research Finds Retrovirus Linked to Schizophrenia, Suggesting Disease Is Not Solely Genetic," *Washington Post,* April 10, 2001; Rick Weiss, "Research Casts Doubt on 'Gay Gene' Theory; Study Finds Nothing within X Chromosome That Predicts Male Homosexuality," *Washington Post,* April 23, 1999; Arthur Allen, "Nature and Nurture: When It Comes to Twins, Sometimes It's Hard to Tell the Two Apart," *Washington Post,* January 11, 1998; Christine Russell, "The Weight Goes On: Genes' Contribution to Obesity Proves to Be a Complex Puzzle," *Washington Post,* February 24, 1998; Sally Squires, "Bed-wetting a Common Inconvenience: Though It May Worry Young Campers and Their Parents, Doctors Say Most Cases Resolve Themselves," *Washington Post,* April 8, 1997.

3. See James K. Glassman, "Who's Afraid of Human Cloning?" *Washington Post,* February 10, 1998.

4. "Elements Through the Ages," http://www.encyclopedia.com/html/section/element_theelementsthroughtheages.asp (Accessed November 16, 2002).

2. SCIENTIfiC FOUNDATIONS

1. See Kenneth J. Burke, "The 'XYY Syndrome': Genetics, Behavior, and the Law," *Denver Law Journal* 46 (1969): 261.

2. See Michael Woods, "Junk DNA Gets New Attention," *Pittsburgh Post Gazette,* March 18, 1994.

3. See George Huntington, *Medical & Surgical Reporter* 26 (1872).

4. See David Nakamura, "A Stacked Deck, but a Spirited Game of Living: Disease Doesn't Stop 6-Year-Old—Family Hopes to Beat the Odds of Cystic Fibrosis," *Washington Post,* April 23, 1998.

5. See David Brown, "Gene For an Inherited Form of Breast Cancer Is Located, Finding Called 'Important'—but Not a Cure," *Washington Post,* September 15, 1994.

6. This and the following description of the early history of the science of genetics is found in Eric S. Lander, "Scientific Commentary: The Scientific Foundational and Medical and Social Prospects of the Human Genome Project," *Journal of Law, Medicine & Ethics* 26 (1998): 184.

7. See Evelyn Fox Keller, *Century of the Gene* (Cambridge, Mass.: Harvard University Press, 2000).

8. For a description of the process of discovery and the early history of the Human Genome Project, see Robert Cook-Deegan, "Origins of the Human Genome Project," http://www.piercelaw.edu/risk/vol5/spring/cookdeeg.htm (Accessed November 16, 2002).

3. FOUR REVOLUTIONS

1. See National Commission on the Future of DNA Evidence, *The Future of Forensic DNA Testing: Predictions of the Research and Development Working Group* (Washington, D.C.: U.S. Dept. of Justice, 2000), pp. 13–20, 46–61.

2. See Robert Goodwin and Jimmy Gurule, *Criminal and Scientific Evidence* (Charlottesville, Va.: Michie, 1997), p. 287.

3. See National Commission on the Future of DNA Evidence, *The Future of Forensic DNA Testing.*

4. 545 N.Y.S.2d 985 (Sup. Ct. 1989).

5. U.S. Department of Justice, Federal Bureau of Investigation, *A Report to the American People on the Work of the FBI 1993–1998,* http://www.totse.com/en/politics/federal_bureau_of_investigation/162451/html (Accessed November 15, 2002).

6. See Ronald Bailey, "Unlocking the Cells," *Reason Magazine,* January 1, 2000, p. 50.

7. See Armed Forces Institute of Pathology, "Repository History," http://www.afip.org/Departments/oafme/dna/history.htm (Accessed November 16, 2002).

8. "A Nation Challenged: US Seeks Source for Some Bin Laden DNA," *New York Times,* February 28, 2002.

9. Hugh McCann, "Military Uses DNA for High-Tech Dogtag," *Detroit News,* March 13, 1995. Cited in Aaron P. Stevens, "Arresting Crime, Expanding the Scope of DNA Databases in America," *Texas Law Review* 79 (2001): 921.

10. Public Law No. 103-322, 108 Stat. 2065 (codified at 42 USC §13701).

11. See Stevens, "Arresting Crime, Expanding the Scope of DNA Databases in America," pp. 926–927.

12. Ibid., p. 946.

13. Ibid., p. 948.

14. In the Matter of the Appeal in Maricopa Cty. Juvenile Action Nos. JV-512600 and JV-512797, 930 P. 2d 496 (Ct. App. Ariz. 1996).

15. "Anti-Terror Compromise," *Boston Globe,* October 4, 2001.

16. Stevenson Swanson, "Some Fear Letting Genetic Genie Out of the Bottle," *Chicago Tribune,* February 2, 1999, cited in Stevens, "Arresting Crime, Expanding the Scope of DNA Databases in America," p. 956.

17. See generally, Joseph Wambaugh, *The Blooding* (New York: Morrow, 1989), pp. 168–69.

18. See Lori Andrews and Dorothy Nelkin, *The Body Bazaar: The Market for Human Tissue in the Biotechnology Age* (New York: Crown Publishers, 2001), pp. 102–103.

19. Kirsten Downey Grimsley, "Bias in a Brave New World: Senate Urged to Pass Laws Barring Genetic Discrimination," *Washington Post,* July 21, 2000.

20. See Barbara Mahany, "Tay Sachs Test Eases the Fears of Orthodox Jews," *Chicago Tribune,* February 7, 1994.

21. See John Pomfret, "In China's Countryside, 'It's a Boy!' Too Often," *Washington Post,* May 29, 2001.

22. Yury Verlinsky et al., "Preimplantation Diagnosis for Early-Onset Alzheimer Disease Caused by V717L Mutation," *Journal of the American Medical Association* 287 (2002): 1018.

23. See "Va. Clinic Develops System for Choosing Sex of Babies," *Washington Post,* September 10, 1998.

24. *Safer* v. *Pack,* 677 A.2d 1188 (NJ Super. Ct. App. Div. 1996).

25. Ohio Revised Code §1751.64.

26. Nicholas Wade, *Life Script: How the Human Genome Discoveries Will Transform Medicine and Enhance Your Health* (New York: Simon & Schuster, 2001), pp. 87–92.

27. Ibid., p. 94.

28. Francis S. Collins, "The Shattuck Lecture—Medical and Societal Consequences of the Human Genome Project," *New England Journal of Medicine* 341 (1999): 28, 30.

29. "Motorola Ships Its First eSensor™ DNA Biochip For Use In Clinical Trials: A New Agreement with a Global Healthcare Company Advances This Cost-effective Detection Technology" (October 16, 2001), http://www.motorola.com/lifesciences/news/mor1oc01.htm (Accessed December 18, 2002).

30. See Marlene Cimons, "How a Body Responds to Drugs Depends on the Genes," *Los Angeles Times,* July 24, 2000.

31. See Byron Spice, "Heart Drug May Counteract Defective Gene: Small Pitt Study Shows Great Promise for Customized Treatments," *Pittsburgh Post Gazette,* March 27, 2001.

32. For a description of the experiment, see Larry Thompson, *Correcting the Code: Inventing the Genetic Cure for the Human Body* (New York: Simon & Schuster, 1994).

33. See Rick Weiss, "Genetic Therapy Apparently Cures 2: French Team's Feat Would Be a First," *Washington Post,* April 28, 2000.

34. See Judith Arden et al., *Law, Science and Medicine* (New York: Foundation Press, 1989), pp. 168–170.

35. The account of Gelsinger's death and its aftermath comes from Sheryl Gay Stolberg, "The Biotech Death of Jesse Gelsinger," *New York Times Sunday Magazine,* November 28, 1999, p. 137.

36. Sheryl Gay Stolberg, "Teenager's Death Is Shaking Up the Field of Human Gene-Therapy Experiments," *New York Times*, Jan. 27, 2000.

37. Lewis Thomas, *The Lives of the Cell* (New York: Viking Press, 1974).

38. "American Association for the Advancement of Science, Human Inheritable Genetic Modifications" (September 2000), http://www.aaas.org/spp/sfrl/germline/report.pdf (Accessed November 15, 2002).

39. See Erik Parens and Eric Juengst, "Inadvertently Crossing the Germ Line," *Science* 292 (2001): 397.

40. See Lori Andrews, "Predicting and Punishing Antisocial Acts: How the Criminal Justice System Might Use Behavioral Genetics," in Ronald A. Carson and Mark A. Rothstein, eds., *Behavioral Genetics* (Baltimore/London: Johns Hopkins University Press, 1999), p. 124.

41. See Charles C. Mann, "Behavioral Genetics in Transition," *Science* 264 (1994): 1686.

42. Ibid.

43. See *People* v. *Tanner,* 91 Cal. Rptr. 656 (Cal. Ct. App. 1970).

44. *People* v. *Yukl,* 372 N.Y.S.2d 313 (N.Y. Sup. Ct. 1975).

45. Associated Press, "Disease Cited in Murder Acquittal," *Cleveland Plain Dealer,* Sept. 29, 1994.

46. See Andrews, "Predicting and Punishing Antisocial Acts," p. 116.

47. See Stadler, "Comment," *Emory Law Journal* 46 (1997): 1285.

48. Allen Buchanan et al., *From Chance to Choice: Genetics and Justice* (New York: Oxford University Press, 2000), p. 27.

49. This quote and the subsequent history of eugenics come from Philip R. Reilly, "Eugenics, Ethics, Sterilization Laws," in Thomas M. Murray and Maxwell J. Mehlman, eds., *Encyclopedia of Ethical, Legal, and Policy Issues in Biotechnology* (New York: John Wiley & Sons, 2000), pp. 204–208.

50. Ibid.; Buchanan et al., *From Chance to Choice*, p. 27.

51. 274 U.S. 200 (1927).

52. These details are found in Paul A. Lombardo, "Three Generations, No Imbeciles: New Light on Buck v. Bell," *N.Y.U. Law Review* 60 (1985): 30, 49–62.

53. The case is *Relf* v. *Weinberger*, 372 F. Supp. 1196 (D. DC 1974), vacated as moot, 565 F.2d 722 (DC Cir. 1977). The background is described in Donna Franklin, "Beyond the Tuskegee Apology," *Washington Post*, May 29, 1997. The United States and Germany are not the only countries to have engaged in large-scale, eugenics-based sterilizations. Between 1935 and 1976, approximately 60,000 people, mostly women, were sterilized in Sweden under a program justified in large part by its promise to reduce the cost of the Swedish welfare system by reducing the number of "inferiors" who would have to be supported. In 1999, the Swedish government agreed to pay reparations to the survivors. See Dan Balz, "Sweden Sterilized Thousands of 'Useless' Citizens for Decades," *Washington Post*, August 29, 1997. An editorial by Martin Bobrow in *Journal of Medical Genetics* 32 (1995): 409 describes China's eugenics program.

54. Peter Breggin, quoted in Mann, "Behavioral Genetics in Transition."

55. For a description of the program and what happened to the twelve children who were born under it, see Douglas A. Blackmon, "A Breed Apart: A Long-Ago Effort to Better the Species Yields Ordinary Folks," *Wall Street Journal*, August 17, 1999.

56. The Repository for Germinal Choice is described in "Owner of 'Genius' Sperm Bank Pleased by the Results," *New York Times*, December 11, 1984.

57. Judge Whitbeck in *Taylor* v. *Kurapati*, 600 NW2d 670 (Mich. Ct App. 1999).

58. Daniel Wickler, "Can We Learn from Eugenics?" *Journal of Medical Ethics*, vol. 25 (1999): 183.

4. THE FIFTH REVOLUTION

1. See Norman Daniels, *Just Health Care* (New York: Cambridge University Press, 1985); Allen Buchanan et al., *From Chance to Choice: Genetics and Justice* (New York: Oxford University Press, 2000).2. See Eric Juengst, "What Does 'Enhancement' Mean?" in Erick Parens, ed., *Enhancing Human Traits: Ethical and Social Implications* (Washington, D.C.: Georgetown University Press, 1998).

3. Kenneth Weiss, "The Egg Brokers," *Los Angeles Times*, May 27, 2001.

4. "Owner of 'Genius' Sperm Bank Pleased by Results," *New York Times*, December 11, 1984.

5. See Dorothy C. Wertz and John C. Fletcher, "Fatal Knowledge? Prenatal Diagnosis and Sex Selection," *Hastings Center Report* 19 (1989): 21; Lynne

Marie Kohm, "Sex Selection Abortion and the Boomerang Effect of a Woman's Right to Choose: A Paradox of Skeptics," *William & Mary Law Review* 4 (1997): 91. For an expansive view of genetic selection of offspring characteristics, including gender, see John A. Robertson, "Genetic Selection of Offspring Characteristics," *Boston University Law Review* 76 (1996): 421.

6. "Cheerleader Case Sentence," *New York Times,* September 10, 1996.

7. Jane Mayer, "It Helps to Thank Headmistress When She Offers Cookies," *Wall Street Journal,* September 29, 1982.

8. American Society of Plastic and Reconstructive Surgeons, "Media Center: Five Year Trends in Cosmetic Surgery 1992 vs. 1997," http://www.plastic surgery.org/mediactr/97change.htm (Accessed May 12, 1998).

9. Nick A. Gapherty, "Performance-Enhancing Drugs," *Sports Medicine* 26 (1995): 433.

10. Daniel A. Smith and Paul J. Perry, "The Efficacy of Ergogenic Agents in Athletic Competition Part II: Other Performance-Enhancing Drugs," *Annals of Pharmacy* 26 (1992): 653.

11. Ibid.

12. See "What Price Perfection," *Orlando Sentinel,* October 12, 1991.

13. Michael Walzer, *Spheres of Justice* (New York: Basic Books, 1983), p. 10.

14. Beth S. Finkelstein et al., "Insurance Coverage, Physician Recommendations, and Access to Emerging Treatments: Growth Hormone Therapy for Childhood Short Stature," *Journal of the American Medical Association* 279 (1998): 663.

15. Richard J. Paulson, "Assisted Reproductive Technologies," *Western Journal of Medicine* 165 (1996): 377.

16. *U.S. News and World Report,* May 13, 1996, p. 20.

17. American Society of Plastic and Reconstructive Surgeons, "1997 Cosmetic Procedures," http://www.plasticsurgery.org/mediactr/97change.htm (Accessed May 12, 1998).

18. Jon W. Gordon, "Genetic Enhancement in Humans," *Science* 283 (1999): 2023, 2024.

5. SAFETY AND EFFECTIVENESS

1. See Robin Herman, "Human Growth Hormone Experiments to Resume; Trials Involve Children Ages 5 to 15," *Washington Post,* June 29, 1993.

2. Baruch A. Brody, *Ethical Issues in Drug Testing, Approval, and Pricing* (New York: Oxford University Press, 1995), p. 23.

3. Paul Berg et al., "Potential Biohazards of Recombinant DNA Molecules," Letter, *Science* 185 (1974): 185.

4. For the story of the Asilomar Conference, see Charles Weiner, "Recombinant DNA Policy: Asilomar Conference," in Thomas M. Murray and Max-

well J. Mehlman, eds., *Encyclopedia of Ethical, Legal, and Policy Issues in Biotechnology* (New York: John Wiley & Sons, 2000), vol. II, p. 910. The conference is described by one of the lawyers who attended in Roger B. Dworkin, "Science, Society, and the Expert Town Meeting: Some Comments on Asilomar," *Southern California Law Review* 51(1978): 1471.

5. See Recombinant DNA Advisory Committee, National Institutes of Health, 42 Fed. Reg. 49596 (1978); Sheldon Krimski, *Genetic Alchemy: The Social History of the Recombinant DNA Controversy* (Boston: MIT Press, 1982).

6. See Gina Kolata, "Little-Known Panel Challenged to Make Quick Cloning Study," *New York Times,* March 18, 1997 ("Dr. Wilmut's feat shocked the world, for even most scientists had assumed that the cloning of adults was biologically impossible and was merely the stuff of science fiction").

7. See Rick Weiss, "Dolly: 'A Sheep in Lamb's Clothing': Clones Inherit Age with Genes, Studies Show," *Washington Post,* May 27, 1999.

8. See Rick Weiss, "Middle-Aged Dolly Develops Arthritis; Questions on Clones' Aging Raised," *Washington Post,* January 5, 2002.

9. See Rick Weiss, "Dolly's Premature Aging Not Evident in Cloned Cows; Finding Hints at Curative Role for Cloned Human Embryos," *Washington Post,* April 28, 2000.

10. See John Schwartz, "Is FDA Too Quick to Clear Drugs? Growing Recalls, Side-Effect Risks Raise Questions," *Washington Post,* March 23, 1999; "$43 Million Order in Drug Suit," *New York Times,* December 22, 2001.

11. See 41 Fed. Reg. 27902 (1976).

12. This and the following paragraphs are adapted from Lori B. Andrews, Maxwell J. Mehlman, and Mark A. Rothstein, *Genetics: Ethics, Law and Policy* (St. Paul, Minn., West Group, 2002), pp. 380–381. For a history of the early RAC, see Krimski, *Genetic Alchemy.* 13. Department of Health and Human Services, National Institutes of Health, Recombinant DNA Research: Request for Public Comment on "Points to Consider in the Design and Submission of Human Somatic-Cell Gene Therapy Protocols," 50 Fed. Reg. 2940 (1985).

14. U.S. Food and Drug Administration, Statement of Policy for Regulating Biotechnology Products, 49 Fed. Reg. 50878 (1984).

15. This was in a notice in the Federal Register, 61 Fed. Reg. 35774 (1996).

16. Federal Food, Drug, and Cosmetic Act, 21 USC § 321(g)(1).

17. 21 CFR §600.3(h).

18. U.S. Food and Drug Administration, Notice of General Surgery and Plastic Surgery Devices Panel Recommendation, 61 Fed. Reg. 58,195–58,196 (1996).

19. FDC, The Gray Sheet, June 22, 1992, p. 11.

20. See Elizabeth C. Price, "Does the FDA Have Authority to Regulate Human Cloning?" *Harvard Journal of Law and Technology* 11 (1998): 619.

21. See e.g., *Chaney* v. *Heckler,* 718 F.2d 1174, 1179 (5th Cir. 1983) (noting that the legislative history of the Federal Food, Drug, and Cosmetic Act demonstrates congressional intent to prohibit the FDA from regulating the practice of medicine). The FDA itself acknowledges this. In 1972, for example, it stated, "It is clear that Congress did not intend [the FDA] to regulate or interfere with the practice of medicine." 37 Fed. Reg. 16,503, 16,504 (1972).

22. Federal law only requires IVF clinics to report their success rates at producing live births. See Federal Fertility Clinic Success Rate and Certification Act of 1992, Pub. L. 102-493, 106 Stat. 3146, October 24, 1992. A few states also regulate IVF clinics. For a description, see New York State Task Force on Life and the Law, Report on Assisted Reproductive Technologies (1997).

23. Stephanie Nano, "Studies Find Test-Tube Babies Run Greater Risk of Low Weight, Defects," *Cleveland Plain Dealer,* March 7, 2002.

24. See *Washington Legal Foundation* v. *Friedman,* 13 F. Supp. 2d 51, 74 (D.DC 1998), amended by 36 F. Supp. 2d 16 (D.DC 1999) and by 36 F. Supp. 2d 418 (D.DC 1999). The act is codified at 21 USC § 360aaa.

25. See Rita Rubin, "Giving Growth a Synthetic Hand: The Use of Hormone Sparks Debate," *Dallas Morning News,* July 7, 1986.

26. "Revoke Oregon's License to Kill," *American Medical News,* September 15, 1997, http://www.ama-assn.org/sci-pubs/amnews/amn_97/edit0915.htm (Accessed December 18, 2002).

27. *Grimes* v. *Kennedy Krieger Institute, Inc.,* 781 A.2d 807 (Md. Ct. App. 2001).

6. AUTONOMY

1. Hippocrates, "Decorum," in *Hippocrates,* trans. W. H. S. Jones, 2nd ed. (Cambridge, Mass.: Harvard University Press, 1967), vol. 2, p. 297, quoted in Judith Areen et al., *Law, Science and Medicine* (New York: Foundation Press, 1984), p. 371.

2. *Lauro* v. *Travelers Insurance Company,* 261 So.2d 261 (La. 1972).

3. *Canterbury* v. *Spence,* 464 F.2d 772 (D.C. Cir. 1972).

4. Charles Lidz et al., *Informed Consent: A Study of Decisionmaking in Psychiatry* (New York: Oxford University Press, 1994), p. 28; Paul S. Appelbaum et al., "False Hopes and Best Data: Consent to Research and the Therapeutic Misconception," *Hastings Center Report* 17 (1987): 20.

5. Thomas M . Murray, "The Ethics of Drugs in Sport," in Richard H. Strauss, ed., *Drugs and Performance in Sports* (Philadelphia: W. B. Saunders, 1987), p. 11.

6. "Courting Danger," *Washington Post,* August 29, 2001.

7. Dinesh D'Souza, "Staying Human: The Danger of Techno-Utopia," *National Review,* January 22, 2001.

8. Mary Powers, "Shaping Their Futures: Teens Turning to Plastic Surgery to Achieve Improved Bodies," *San Diego Union,* September 23, 2000.

9. 21 CFR §50.3.

7. AUTHENTICITY

1. See International Olympic Committee, "IOC's Fight against Doping: A Brief History," http://www.olympic.org/uk/organisation/commissions/medical/index_uk.asp (Accessed November 16, 2002).

2. "'The Drug Games': The legacy Sydney doesn't want," http://cbs.Sportsline.com/u/ce/multi/0,1329,2859112_15,00.html (Accessed December 18, 2002).

3. Duncan Mackay, "Olympic Games: EPO Use Hit New Heights in Salt Lake," *The Guardian*, April 10, 2002.

4. See Olga Connolly, "Steroid Debate: 'Enhanced' vs. Natural Athletes," *Washington Post*, September 13, 1988.

8. ACCESS

1. See W. Hughes, *Alexander Fleming and Penicillin* (London: Priority Press, 1974), pp. 59–79.

2. See James Childress, "Triage in Neonatal Intensive Care: The Limitations of a Metaphor," *Virginia Law Review* 69 (1983): 547, 551–552.

3. Beth S. Finkelstein et al., "Insurance Coverage, Physician Recommendations, and Access to Emerging Treatments: Therapy for Childhood Short Stature," *Journal of the American Medical Association* 279 (1998): 663.

4. *Report of the New York State Task Force on Life and the Law on Assisted Reproductive Technologies* (New York: New York State Dept. of Health, 1998), p. 165.

5. *Diamond* v. *Chakrabarty,* 447 US 303 (1980).

6. *Amgen, Inc.* v. *Chugai Pharmaceutical Co., Ltd.,* 927 F.2d 1200 (Fed. Cir. 1991).

7. See U.S. Department of Energy, Office of Biological and Environmental Research, "Genetics and Patenting," http://www.ornl.gov/hgmis/elsi/patents.html (Accessed November 16, 2002).

8. Nicholas Thompson, "Gene Blues: Is the Patent Office Prepared to Deal with the Genomic Revolution?" *Washington Monthly,* April 2001.

9. U.S. Census Bureau, *Money Income in the United States 2000,* http://www.census.gov/prod/2002pubs/p60-219.pdf (Accessed November 16, 2002).

10. See 42 USC § 1395y(a)(1)(A).

11. See, e.g., Mass. Gen. Laws Ann. Ch. 118G, section 1 ("medically necessary services shall not include . . . cosmetic surgery").

12. *Viveros* v. *Dep't of Health & Welfare*, 889 P.2d 1104 (Idaho 1995).

13. See e.g., Cal. Ins. Code section 10119.6 (St. Paul, Minn.: West Group, 1993).

14. The federal law is the Employee Retirement Income Security Act, 29 USC § 1144.

15. The figure is based on an average cost of $30,000 per fertilization for the approximately 4 million live births in the U.S. every year. The figure excludes unsuccessful pregnancies that also might have involved in vitro fertilization (IVF) and attempts at genetic enhancement. Medicare and federal budget amounts are based on U.S. government statistics for 2000, which can be found at http://cms.hhs.gov/researchers/pubs/datacompendium/ (Accessed November 18, 2002).

9. INEQUALITY AND UNFAIRNESS

1. Lori Andrews, personal communication; see also, e.g., http://www. xent.com/FoRK-archive/fal196/0415.html (Accessed November 15, 2002).

2. John Rawls, *A Theory of Justice* (Cambridge, Mass.: Harvard University Press, 1971), p. 104.

3. Ronald Dworkin, "What Is Equality, Part 2: Equality of Resources," in *Philosophy and Public Affairs* 10 (1981): 185, 293.

4. Robert Nozick, *Anarchy, State and Utopia* (New York: Basic Books, 1974), pp. 151–52.

5. Eric Rakowski, *Equal Justice* (Oxford: Clarendon, 1991), p. 159.

6. John Gardner, *Excellence, Can We Be Equal and Excellent Too?* (New York: Harper, 1984), p. 30.

7. Rawls, *A Theory of Justice*, p. 78.

8. Thomas Nagel, *Equality and Partiality* (New York: Oxford University Press, 1991), p. 121.

9. Rawls, *A Theory of Justice*, pp. 534–541.

10. Dworkin, "What Is Equality, Part 2," pp. 185, 285.

11. Michael Walzer, *Spheres of Justice* (New York: Basic Books, 1983), p. xiii.

12. Frank Parkin, *Class Inequality and Political Order: Social Stratification in Capitalist and Communist Societies* (New York: Praeger Publishers, 1971), p. 48.

13. David B. Grusky and Azumi Ann Takata, "Social Stratification," in Edgar F. Borgatta and Marie L. Borgatta, eds., *Encyclopedia of Sociology* (New York: Macmillan Reference USA, 1992), p. 1965. (This and portions of the nearby text were originally published in Maxwell J. Mehlman, "The Law of

Above Averages: Leveling the New Genetic Enhancement Playing Field," *Iowa Law Review* 85 [2000]: 517, 550–552.)

14. John H. Schaar, *Legitimacy in the Modern State* (New York: New York University Press, 1981), p. 195.

15. *The Politics of Aristotle,* trans. Peter L. Phillips Simpson (Chapel Hill: University of North Carolina Press, 1997), paras. 1295b34, 1296a7.

16. Kenneth Karst, "Symposium: Voices of the People: Essays on Constitutional Democracy in Memory of Professor Julian N. Eule: Participation and Hope," *UCLA Law Review* 45 (1998): 1773.

17. Ronald M. Glassman, *The Middle Class and Democracy in Socio-Historical Perspective* (Leiden, the Netherlands: E. J. Brill, 1995), p. 216.

18. "Caste Violence Flares Up Again; Of 34 Dead, Most Are Women, Children, and Old Men," *New York Times,* June 18, 2000.

10. HUBRIS

1. *Orange County (California) Register,* July 25, 1995.

2. *Boston Globe,* May 14, 1995.

3. Jonah Goldberg, "Frankenjournalists," http://www.nationalreview.com/goldberg/goldberg072301.shtml (Accessed November 15, 2002).

4. C. B. Fehilly et al., "Interspecific Chimaerism between Sheep and Goat," *Nature* 307 (1984): 634.

5. K. T. Paige et al., "Injectable Cartilage," *Plastic and Reconstructive Surgery* 96 (1995): 1390.

11. SOLUTIONS

1. See Alex Kucynski, "In Quest for Wrinkle-Free Future, Frown Becomes a Thing of the Past," *New York Times,* February 7, 2002.

2. "FDA Approves New Injectable Product for Incontinence," http://www.fda.gov/bbs/topics/ANSWERS/ANS00531.html (Accessed November 15, 2002).

3. Anabolic Steroids Control Act, Pub. L. 101-647, 1901-1907, 104 Stat. 4851 (1990).

4. 21 USC § 812.

5. 21 USC § 333(e)(1).

6. *United States* v. *Rutherford,* 442 U.S. 544 (1979).

7. *Cleveland Board of Education* v. *LaFleur,* 414 U.S. 632 (1974).

8. *Moore* v. *City of East Cleveland,* 431 U.S. 494 (1977).

9. *Halderman* v. *Pennhurst State School & Hospital,* 707 F.2d 702 (3d Cir. 1983).

10. *Wisconsin* v. *Yoder,* 406 U.S. 205, 213–214 (1972); *Pierce* v. *Society of Sisters,* 268 U.S. 510 (1925).

11. *Parham* v. *J.R.,* 442 U.S. 584, 603 (1979).

12. See, e.g., *Perry* v. *City of Norfolk,* 194 F.2d 1305 (4th Cir. 1999); *Thomason* v. *SCAN Volunteer Services, Inc.,* 85 F.3d 1365 (8th Cir. 1996); *F.K.* v. *Iowa District Court for Polk County,* 630 N.W.2d 801 (Iowa 2001); *Hooper* v. *Rockwell,* 513 S.E.2d 358 (S.C. 1999).

13. *Murray* v. *State of Florida,* 384 F. Supp. 574 (S.D. Fla. 1974); *State* v. *DuFresne,* 782 So.2d 888 (Ct. App. Fla. 2001).

14. *Florida* v. *McDonald,* 785 So. 2d 640 (Ct. App. Fla. 2001).

15. *Halderman* v. *Pennhurst State School and Hospital,* 707 F.2d 702, 709 (3rd Cir. 1983).

16. 316 US 535, 541 (1942).

17. *Planned Parenthood* v. *Casey,* 505U.S. 833, 857 (1992).

18. *Wisconsin* v. *Yoder,* 406 U.S. 205, 215 (1971).

19. *Jacobellis* v. *Ohio,* 378 U.S. 184 (1964).

20. *Tennessee* v. *Gardner,* 105 S. Ct. 1694 (1985); *Regan* v. *Time, Inc.,* 104 S. Ct. 3262 (1984); *Buckley* v. *Valeo,* 96 S. Ct. 612 (1976).

21. *Lifchez* v. *Hartigan,* 735 F. Supp. 1361 (N.D. Ill. 1990).

22. *Cameron* v. *Board of Education,* 795 F. Supp. 228 (S.D. Ohio 1991).

23. See, e.g., *Elliot* v. *North Carolina Psychology Board,* 485 S.E. 2d 882 (S.C. 1997).

24. "Report on Ethical Issues Related to Prenatal Genetic Screening," *Archives of Family Medicine* 3 (1994): 633, 637–39.

25. Alex Kucynski, "In Quest for Wrinkle-Free Future, Frown Becomes a Thing of the Past," *New York Times,* February 7, 2002.

26. Dietary Supplement Health and Education Act, Public Law No. 103-417, 108 Stat. 4325 (1994).

27. Mark Ziegler, "Illegal Doping Is Everywhere Now, and the Culprits Are Rarely Caught," *San Diego Union-Tribune,* August 17, 1997.

28. See Zad Leavy and Jerome M. Kummer, "Criminal Abortion: Human Hardship and Unyielding Laws," *Southern California Law Review* 35 (1962): 123.

29. Dan Weikel, "Prescription Fraud: Abusing the System," *Los Angeles Times,* August 18, 1996; Charles W. Hall, "A Prescription for Trouble in Suburbia: Virginia Police Step Up Efforts against Fraud at Pharmacy," *Washington Post,* May 26, 1996 (using DEA statistics).

30. See RAC Guidelines for Research Involving Recombinant DNA, III-c-1, app. M., http://www.od.nih.gov/oba/rac/guidelines_02/nih_guidelines_apr_02.htm#_toc7261565 (Accessed November 16, 2002).

31. For a general discussion of research moratoria, see Robert Cook-Deegan, "Cloning Human Beings: Do Research Moratoria Work?" in *Cloning Human Beings: Report and Recommendations of the National Bioethics Advisory Commission* (1997, see http://www.georgetown.edu/research/nrcbl/nbac/pubs/cloning2/ccb.pdf [Accessed November 16, 2002]), vol. 2, p. H8.

32. This term may have originated in an article by Mayor Federico in the *UNESCO Courier,* "The Responsibility of Scientists: Science and Society, Part 2," May 1997.

33. See Jon S. Batterman, "Brother Can You Spare a Drug?" *Hofstra Law Review* 19 (1990): 191, 207–208.34. Liz Brody, "Easy Drugs OnLine," *Glamour,* April 2000, p. 102.

35. "U.S. Customs—Traveler Information," http://www.customs.ustreas. gov/travel/travel.htm (Accessed November 16, 2002).

36. Edward Cody, "AIDS Victims Seek Drugs in Mexico: FDA Discounts Effect of Last-Chance Cures Barred in U.S.," *Washington Post,* September 10, 1986.

37. "Resolution on the Protection of Human Rights and Dignity with Regard to the Application of Biology and Medicine," *Official Journal of the European Communities,* no. C320 (1996).

38. "Team to Attempt Human Cloning" (March 9, 2001), http://www. cnn.com/2001/WORLD/europe/03/09/clone/index.html (Accessed December 18, 2002).

39. Richard Saltus, "Are We Getting Closer to Cloning Human Beings?" *Boston Globe,* September 5, 2000.

40. See U.S. Department of Defense, "Proliferation: Threat and Response," http://www.defenselink.mil/pubs/prolif/me_na.html (Accessed December 18, 2002).

41. Paul Grecian, "British Agent Facing U.S. Arms Charges," *New York Times,* January 26, 1996.

42. Judy Mann, "Money Spent on Drug War Could Be Put to Better Use," *Washington Post,* October 17, 2001, p. C12; U.S. Department of Justice, Drug Enforcement Administration, "Inside the DEA, DEA Staffing & Budget," http://www.usdoj.gov/dea/agency/staffing.htm (Accessed November 16, 2002).

43. This is the title of an article on the control of offshore access to genetic enhancements written by the author and Kirsten Rabe and published in 2002 in the *American Journal of Law and Medicine.*

44. See 19 USC § 482; *US* v. *Himmelwright,* 551 F.2d 991 (5th Cir. 1977).

45. 42 USC § 264. The diseases are listed by executive order. See, e.g., Executive Order no. 12,452, 48 Fed. Reg. 56927 (1983).

46. See *Zemel* v. *Rusk,* 381 US 1 (1965) (upholding restrictions on travel to Cuba); 22 USC § 7209.

47. 8 USC § 1185.

48. See, e.g., *US* v. *Bowman,* 260 US 94 (1922) (conspiracy to defraud).

49. *State* v. *Harvey,* 2 F.3d 1318 (3d Cir. 1993).

50. See generally Ethan Avram Nadelmann, *Cops across Borders: The Internationalization of U.S. Criminal Law Enforcement* (University Park: Pennsylvania State University Press, 1997).

51. See M. Christina Ramirez, "The Balance of Interests between National Security Controls and First Amendment Interests in Academic Freedom," *Journal of College and University Law,* vol. 13 (1986): 179.

52. 31 USC § 5314.

53. 26 USC § 7206(1); *US* v. *Clines,* 958 F.2d 578 (4th Cir. 1992).

54. See "The Universal Declaration of Human Rights"; the "Declaration on the Elimination of Violence against Women," 33 I.L.M. 1049 (1994); and the "Declaration of the Rights of the Child" (1959).

55. See "Beyond Cloning: Protecting Humanity from Species-Altering Experiments," http://www.bumc.bu.edu/www/sph/lw/website/index.htm (Accessed November 16, 2002).

56. 50 USC § 1701.

57. 50 USC § 1702(a).

58. See Adam Smith, "A High Price to Pay: The Costs of the U.S. Economic Sanctions Policy and the Need for Process Oriented Reform," *UCLA Journal of International Law and Foreign Affairs* 4 (Fall/Winter 1999–2000): 325, 327.

59. 2001 Cong. US SJ 23 (Joint Resolution S.J. Resolution 23).

60. 50 USC § 1541.

61. Public Law No. 107-56, 115 Stat. 272 (2001).

12. BETTER SOLUTIONS

1. Robert Nozick, *Anarchy, State, and Utopia* (New York: Basic Books, 1974), p. 371, n.41, quoting Isaiah Berlin, "Equality," in Frederick A. Olafson, ed., *Justice and Social Policy* (Upper Saddle River, N.J.: Prentice Hall, 1961), p. 131.

2. A more elaborate and elegant description of the dispute between welfare and resource egalitarians can be found in Ronald Dworkin, "What Is Equality? Part I: Equality of Welfare" and "Part II: Equality of Resources," *Philosophy and Public Affairs* 10 (1981): 185, 283.

3. These philosophers include Thomas Nagel (*Equality and Partiality* [New York: Oxford University Press, 1991], pp. 136–137); Madison Powers ("Forget about Equality," *Kennedy Institute of Ethics Journal* 6 [1996]: 136); and Norman Daniels (*Just Health Care* [New York: Cambridge University Press, 1986], p. 28).

4. John Rawls, *A Theory of Justice* (Cambridge, Mass.: Harvard University Press, 1971), pp. 17, 136.

5. *Regents of the University of California* v. *Bakke,* 438 US 265, 297 (1978).

6. Rawls, *A Theory of Justice,* p. 100.

7. 29 CFR § 1630.2(1).

8. *Bragdon* v. *Abbott,* 524 US 624 (1998).

9. Anita Silvers, "A Fatal Attraction to Normalizing," in Erik Parens, ed., *Enhancing Human Traits* (Washington, D.C.: Georgetown, University Press, 1998), p. 121.

10. Harlan Lane et al., *A Journey into the Deaf-World* (San Diego: Dawn Sign Press, 1996), p. 403.

11. Andrew Solomon, "Defiantly Deaf," *New York Times Sunday Magazine,* August 28, 1994, p. 4 (quoting Patty Ladd, a British Deaf scholar).

12. This idea was first proposed by the author in a book he co-wrote with Jeffrey Botkin in 1998 entitled *Access to the Genome: The Challenge to Equality* (Washington, D.C.: Georgetown University Press), and was reiterated in an article entitled "The Law of Above Averages: Leveling the New Genetic Enhancement Playing Field," *Iowa Law Review* 85 (2000). Some of the text is adapted from the former.

13. Ronald P. Keeven, "Pros and Cons of Gambling Amendment: Money Used for Legal Betting Drains Resources of the Poor," *St. Louis Post-Dispatch,* March 27, 1994.

14. Ibid.

15. This and the following historical observations are from Barbara Goodwin, *Justice by Lottery* (Chicago: University of Chicago Press, 1992).

16. *Holmes* v. *New York City Housing Authority,* 398 F.2d 262 (2d Cir. 1968); *Hornsby* v. *Allen,* 330 F.2d 55 (5th Cir. 1964).

17. Deborah S. Pinkney, "Firm Faces Legal Flak over Drug Monitoring Rules," *American Medical News,* November 2, 1990, p. 33.

18. Susan K. Miller, "MS Drug Shortage Prompts Patient Lottery," *New Scientist* 139 (1980): 8.

19. *US* v. *Holmes,* 26 F. Cas. 360 (E.D. Pa. 1842)(No. 15,383).

20. Stephanie Nano, "Studies Find Test-Tube Babies Run Greater Risk of Low Weight, Defects," *Cleveland Plain Dealer,* March 7, 2002.

21. Eric Juengst, personal communication, remarks attributed by Eric Juengst to Bernie Gert.

22. *Relf* v. *Weinberger,* 372 F. Supp. 1196 (D.D.C. 1974), vacated as moot, 565 F.2d 722 (D.C. Cir. 1977).

23. Ibid.

24. Leon Kass, personal communication, March 17, 2001.

25. Diane M. Gianelli, " New York panel urges stricter controls over fertility clinics," *American Medical News,* http://www.ama-assn.org/sci-pubs/amnews/pick_98/pick0518.htm (Accessed December 18, 2002).

26. 100 US 303 (1879) (emphasis added).

27. Insider Trading and Securities Fraud Enforcement Act of 1988, Public Law No. 100-704, 102 Stat. 4677 (1988); Insider Trading Sanctions Act of 1984, Public Law No. 98-376, 98 Stat. 1264 (1984).

28. Restatement (Second) of the Law of Contracts, sec. 161(d) (1981).

29. *Ollerman* v. *O'Rourke Co.*, 288 NW 95 (Wis. 1980).

30. Diagram Group, in Jack Wilkinson, ed., *Rules of the Game* (New York: Paddington Press Ltd., 1974), p. 403.

31. Blakney Boggs, "Your Game Handicaps Help Promote Equal Competition," *Orange Country Register*, August 13, 1998.

32. National Association of Securities Dealers, *NASD Manual* and *Notices to Members*, http://www.nasdr.com/pdf-text/nasd_manual.pdf (Accessed December 18, 2002).

33. Tom Verducci, "Kids' Stuff," *Sports Illustrated*, April 4, 1994, p. 50.

34. Elizabeth Comte, "WBL: A Short Circuit with a Worldwide Reach," *Sporting News*, May 21, 1990, p. 44.

35. "Chess: Drug Testing Has Arrived," *New York Times*, November 16, 1999.

36. John Gardner, *Excellence: Can We Be Equal and Excellent Too?* (New York: Harper, 1984), pp. 109–110.

37. Kurt Vonnegut, "Harrison Bergeron," in *Welcome to the Monkey House* (New York: Dell, 1968), p. 7.

38. Nagel, *Equality and Partiality*, p. 138.

39. The rabbit was created at the request of Chicago artist Eduardo Kac. See "Artist Splices Jellyfish With Rabbit To Make Art," http://www.ekac.org/win.html (Accessed November 16, 2002).

13. SPOTTING ENHANCEMENT

1. Columbia Pictures, 1997.

2. Michael Pollan, "Playing God in the Garden," *New York Times Sunday Magazine*, October 25, 1998, p. 44.

14. CONCLUSION

1. Jon W. Gordon, "Genetic Enhancement in Humans," *Science* 2023 (1999): 283.

Index

Page numbers in italics refer to illustrations.

MAXWELL J. MEHLMAN is Arthur E. Petersilge Professor of Law and Director of the Law-Medicine Center, Case Western Reserve University School of Law, and Professor of Biomedical Ethics at the Case Western Reserve University School of Medicine. He is the co-author of *Access to the Genome: The Challenge to Equality* and co-editor of the *Encyclopedia of Ethical, Legal, and Policy Issues in Biotechnology.*

MEDICAL ETHICS SERIES

David H. Smith and Robert M. Veatch, Editors